HUMAN SPACE FLIGHTS

1957 to the Present Day

Including the Previously Hidden and Secret Flights

John Elliott

First Edition

Editor:
Emily Andsager
Deb A. Watts-Elliott
Jaden Yaboz

Formatting and Cover designed:
Linda Daly

ISBN: 9798879918366

Printed in the United States of America

www.aepublisher.com

Dedication

This book is dedicated to all of those involved in human spaceflight endeavors, from the very early days, all the way into the future.

Preface

There will be those who have a rather difficult time absorbing the information on the following pages and perhaps a few may even suggest that it is utter nonsense. But, no matter what the naysayers say, what you are about to read is all true.

Nikita Khrushchev who was in office until 1964, and now Vladimir Putin, both depraved and iniquitous dictators, denied all of what follows. They hid the truth from the world, erasing the names of all those fallen cosmonauts in order to appear far more technically advanced than the United States. But now the truth is being told, and they both have been proven wrong.

The point here is that the former Soviet Union under Khrushchev, sent up at least fifteen spacecraft carrying seventeen cosmonauts to their deaths, all while trying desperately to beat the United States during what was previously referred to as "The Space Race." And, although it cannot yet be unequivocally proven, there may well have been four more human-occupied spaceflight attempts by the Soviets, all ending in utter tragedies.

On the following pages all seventeen fallen cosmonauts are identified by name, as are their space flights, and dates. It is the author's hope that it will bring some measure of comfort to the remaining family members who may have survived the previous Soviet era purges.

Buckle your seat belts, because what follows is both eye-opening and astounding! It has never been in print before and this is the very first book to ever be published that actually reveals the truth. Every human-occupied spaceflight from the early days until now, including all of the forty women and men who tragically perished during spaceflight attempts, is finally revealed.

Table of Contents

This is truly a series of Earth-shattering events. Never before revealed, they finally bring to light the horror perpetrated by both China and the former Soviet Union during their failed attempts to better the United States in human space flights.

The Flights

During an astonishing eye-opening personal interview with Sergei Khrushchev, the son of Nikita, at Brown University in Providence, Rhode Island over thirty years ago, a number of history-making statements were made. Sergei was an aerospace engineer in the former Soviet Union before moving to the United States, where he became a professor at that prestigious university. And he was one of several people who knew the truth, and was willing to share it, so that the history books could be re-written.

He confirmed that Yuri Gagarin was in fact not the first person to ride up into space as we've all been led to believe, but had been preceded by well over a dozen others, all but one losing their lives because of the Soviet Union's frantic and rushed attempts to beat the United States! And the one who actually did survive before Gagarin went up into orbit? Well, he landed not back in the Soviet Union, but in the northern steppes of China, where he was taken to a hospital for all of his injuries, and then imprisoned for a full year by the Chinese.

And of those injuries? They were the reason why Gagarin parachuted out of his capsule as it descended towards the ground, and didn't ride it all the way down. He knew it was about to crash.

Oh, and about the Chinese, for the first time we will learn that they also lost some of their astronauts (their "taikonauts"). At least three of them perished as the Chinese rushed to put humans up into orbit. The Russians and the Chinese have kept all of this a secret for well over half a century, but now the truth is finally revealed.

What follows is the actual order for all of human space flight, from the very beginning in the fall of 1957 to the present day. This is the very first time that virtually every flight has been listed, including all the previously denied Soviet and Chinese space flights. You won't find this written anywhere else, as much of what follows was deemed to be secret and hidden. But, finally, after well over half of a century, the truth is revealed.

The space flights where men and women perished in their attempts are in bold font. They are as follows:

11/15/1957 … R-5A … Aleksei Ledovsky … he reached an altitude of 200 miles, but died during the sub-orbital flight out of Kapustin Yar … Ledovsky should have made it into the history books, as he was the very first person to go into space, but, under the Soviet Union's paranoia, that honor was taken away from him and his family, as he was also the first person to die there. The flight duration was believed to be 32 minutes, 17 seconds … (USSR).

02/08/1958 … R-5A … Serenti Shiborin … he died during a sub-orbital flight out of Kapustin Yar. The flight duration was indicated as 19 minutes, 09 seconds … (USSR).

01/02/1959 … R-5A … Andrei Mitkov … he was killed during a sub-orbital flight out of Kapustin Yar. The duration was reported to be 22 minutes, 27 seconds … (USSR).

12/11/1959 … experimental winged VKA Myasishchev space plane (Burya) … Mirya (Mira) Gromova … the first female in space for a sub-orbital flight, died during the re-entry attempt. Again, Ms. Gromova should have been in the history books as she was the first woman in space, but, of course, the first female to die there. The total flight duration was reported as just 9 minutes, 44 seconds above a 50 mile altitude before the craft broke apart and burned up in the upper reaches of the atmosphere. That duration time, however, has never been confirmed with 100% certainty, and it could have been just slightly longer, perhaps above the ten minute mark. Nevertheless, Mirya Gromova deserves her rightful place in history … (USSR).

05/16/1960 … Vostok capsule … Vasilli Zavadovsky … he died during an orbital flight out of the Baikonur space center when his capsule burned up during a re-entry attempt. The flight duration is not known … (USSR).

09/27/1960 … Vostok capsule … Ivan Kachur … he was killed during an orbital flight attempt out of the Baikonur space center when there was a significant malfunction during liftoff. The total duration

was reported variously as anywhere from 27 seconds to 3 minutes and 11 seconds ... (USSR).

10/11/1960 ... Vostok capsule ... Piotr Dolgov ... he died during an orbital flight attempt out of the Baikonur space center ... the rocket blew up during launch. The duration was reported to be 44 seconds ... (USSR).

11/28/1960 ... Vostok capsule ... Alexis Graciov ... he died during an orbital flight out of the Baikonur space center, and thought to have left Earth orbit and unable to return. There was, for obvious reasons, no flight duration time given ... (USSR).

02/17/1961 ... modified Vostok capsule ... Gennady Mikhailov ... he died during an orbital flight out of the Baikonur space center after attempting a trans-lunar flight. There were no flight duration times given ... (USSR).

04/07/1961 ... Vostok capsule (Rossiya) ... Vladimir Ilyushin ... the first person to actually survive human space flight, the craft was brought down to Earth, but way off course and landed in China. Suffering injuries due to the bone-jarring landing, Ilyushin was first taken to a hospital for treatment, but then held in a Chinese jail for a full year before his eventual release back to Russia. The flight duration was never reported, but believed to be (approximately) 5 hours and 77 seconds ... (USSR).

04/12/1961 ... Vostok 1 (Kedr (Cedar)) ... Yuri Gagarin ... 1 orbit ... 1 hour, 48 minute duration; due to the reports regarding Vladimir Ilyushin's injuries upon landing in China, Gagarin chose to eject out of the Vostok capsule at approximately 26,000 feet, and landed by parachute ... (USSR).

05/05/1961 ... Mercury-Redstone 3 (Freedom 7) ... Alan Sheppard ... sub-orbital flight. America's first space flight ... 15 minute, 22 second duration ... (USA).

05/16/1961 ... Vostok capsule ... Ludmilla Tokovy and Nikoly (Anatoly) Tokovy ... a married couple, sent up into orbit in order to prove to the world that the Soviet Union had advanced far beyond the

United States, and that, as a married couple, they (according to two former Soviet sources) would engage in sexual intimacy in orbital weightlessness ... the craft apparently spun out of control soon after orbital insertion, and burned up in the atmosphere during re-entry ... (USSR).

06/12/1961 ... Dongfeng 1 (SS-2) medium range missile ... Kuo Yi ... He was launched in the nosecone of this missile, in a standing position, and it was launched to achieve sub-orbital flight ... the reports indicate that it achieve a maximum altitude of 62.3 miles, before burning up in the upper reaches of the atmosphere on its return, the parachute system failing to function ... the flight duration was believed to be 26 minutes and 7 seconds before the nosecone came crashing to the ground ... (China).

07/21/1961 ... Mercury-Redstone 4 (Liberty Bell 7) ... Virgil Grissom ... sub-orbital flight ... 15 minute, 27 second duration. The escape hatch blew off once the spacecraft splashed down, and Grissom jumped out, very nearly drowning ... (USA).

08/06/1961 ... Vostok 2 ... Gherman Titov ... 17 orbits ... 1 day, 01 hour, 18 minute duration ... first flight (survived) of more than 24 hours ... (USSR).

10/14/1961 ... modified Vostok capsule ... Ivan Grachov ... launched from the Baikonur space center, it was intended as a lunar trajectory flight with one loop of the Moon. The capsule failed to swing around the Moon, however, and was lost in deep space. The duration for this flight can not be known, as there was no return to Earth ... (USSR).

02/20/1962 ... Mercury-Atlas 6 (Friendship 7) ... John Glenn ... 3 orbits ... 4 hour, 55 minute duration ... first orbital flight for Americans. Telemetry falsely indicated that the heat shield may have unlatched ... (USA).

05/15/1962 ... Vostok capsule ... Alexis Belokonyov (Belokoniov) ... he was killed during an orbital flight out of the Baikonur space center.

He died during the attempted re-entry when, according to reports, the capsule's initial plunge into the upper atmosphere was misjudged and too steep, causing the craft to bounce off the atmosphere back into deep space where it was non-recoverable. The total duration for the flight is not known … (USSR).

05/24/1962 … Mercury-Atlas 7 (Aurora 7) … Scott Carpenter … 3 orbits … 4 hour, 56 minute duration. Initiated space flight experiments. A manual retrofire error caused a 250-mile landing overshoot … (USA).

07/17/1962 … X-15 Flight 62 … Robert M. White … Achieved a sub-orbital altitude of 59.6 miles (95.9 kilometers), and a speed of 3,831 mph (6,165 km/h) … (USA).

08/11/1962 …Vostok 3 … Andrian Nikolayev ... First "twin" flight along with Vostok 4. Flight duration was 3 days, 22 hours, 22 minutes … (USSR).

08/12/1962 … Vostok 4 … Pavel Popovich ... On the first orbit, Vostok 4 came within three miles of Vostok 3. Flight duration was 2 days, 22 hours, 57 minutes … (USSR).

10/03/1962 … Mercury-Atlas 8 (Sigma 7) … Walter Schirra … Developed techniques for long duration missions (six orbits). It splashed down the closest to its desired target location, just 4.5 miles from the target. Duration was 9 hours, 13 minutes … (USA).

01/17/1963 … X-15 Flight 77 … Joe Walker ... Achieved a sub-orbital altitude of 51.4 miles (82.7 kilometers), and a speed of 3,677 mph (5,918 km/h) … (USA).

05/15/1963 … Mercury-Atlas 9 (Faith 7) … Gordon Cooper ... The first U.S. evaluation of the effects of one full day in space (22 orbits). Performed a manual reentry after a systems failure, and landed four miles from the target. Duration was 1 day, 10 hours, 20 minutes … (USA).

06/14/1963 ... Vostok 5 ... Valeri Bykovsky ... The second twinned flight (with Vostok 6). Duration was 4 days, 23 hours, 08 minutes ... (USSR).

06/16/1963 ... Vostok 6 (Zarya (Dawn)) ... Valentina Tereshkova ... Tereshkova's individual call sign was Chaika (Seagull). First female to survive in space. Came within three miles of Vostok 5. Duration was 2 days, 22 hours, 51 minutes ... (USSR).

06/27/1963 ... X-15 Flight 87 ... Robert Rushworth ... A sub-orbital flight of 53.9 miles (86.7 kilometers), and a speed of 3,425 mph (5,512 km/h) ... (USA).

07/19/1963 ... X-15 Flight 90 ... Joseph A. Walker ... A sub-orbital flight of 65.8 miles (105.9 kilometers), and a speed of 3,710 mph (5,970 km/h), Duration was 11 minutes, 24 seconds ... (USA).

08/22/1963 ... X-15 Flight 91 ... Joseph A. Walker ... A sub-orbital flight of 67.0 miles (107.8 kilometers), and a speed of 3,794 mph (6,106 km/h), Duration was 11 minutes, 09 seconds ... (USA).

10/12/1964 ... Voskhod 1 ... Vladimir Komarov (Ruby 1), Konstantin Feoktistov (Ruby 2), Boris Yegorov (Ruby 3) ... The Voskhod capsule was simply a modified Vostok capsule, in this case stripped down in the interior to accommodate the three cosmonauts, primarily to gain a perceived upper hand on the U.S. Because of the cramped quarters, they could not wear space suits. The Russians only flew the Voskhod type one other time, but never again with three on board. Duration was 1 day, 17 minutes ... (USSR).

03/18/1965 ... Voskhod 2 ... Alexei Leonov (Diamond 2), Pavel Belyayev (Diamond 1) ... The second modified Vostok capsule. This was the first EVA, performed by Leonov using an inflatable airlock, but nearly a disaster, as his spacesuit over inflated with air, and he had an extremely difficult time reentering the capsule once the ten minute EVA was over. Duration was 1 day, 02 hours, 02 minutes ... As an aside, Leonov died at age 85 on October the 11th, 2019 ... (USSR).

In the March to April time frame of the following year (1966), rumors pointed to the possibility of a Voskhod 3 flight planned for a 19 day duration test. It was to carry Boris Volynov and Georgi Shonin into orbit, but, Soviet technicians were never able to provide enough breathable oxygen beyond what would have been available for a 16 day duration flight. Even so, and as hard as it is to believe, the mission very nearly was launched! Cooler heads prevailed, thank God, and Voskhod 3 was canceled at the last minute in favor of the upcoming Soyuz test flights.

03/23/1965 ... Gemini 3 (Molly Brown) ... Virgil Grissom, John Young ... The first American two man crew; the first piloted spacecraft to change its orbital path; the first computer allowing on board calculations of maneuvers. Duration was 4 hours, 53 minutes ... (USA).

06/03/1965 ... Gemini 4 ... James McDivitt, Edward White ... The first American EVA by White (21 minutes); the first American four (4) day flight; a manual reentry was made following a computer failure. Duration was 4 days, 01 hours, 56 minutes ... (USA).

06/29/1965 ... X-15 Flight 138 ... Joe II. Engle ... A sub-orbital flight of 53.1 miles (85.5 kilometers), and a speed of 3,431 mph (5,522 km/h) ... Duration was 10 minutes, 32.3 seconds ... (USA).

08/10/1965 ... X-15 Flight 143 ... Joe H. Engle ... A sub-orbital flight of 51.3 miles (82.6 kilometers), and a speed of 3,549 mph (5712 km/h) ... Duration was 9 minutes, 51.8 seconds ... (USA).

08/21/1965 ... Gemini 5 ... Gordon Cooper, Charles Conrad ... The first use of fuel cells for electrical power; evaluated guidance and navigation systems. Duration was 7 days, 22 hours, 55 minutes ... (USA).

09/28/1965 ... X-15 Flight 150 ... John B. McKay ... A sub-orbital flight of 55.9 miles (90.0 kilometers), and a speed of 3,731 mph (6,004 km/h) ... Duration was 11 minutes, 56.8 seconds ... (USA).

10/14/1965 ... X-15 Flight 153 ... Joe H. Engle ... A sub-orbital flight of 50.4 miles (81.8 kilometers), and a speed of 3,554 mph (5,720 km/h) ... Duration was 9 minutes, 17.7 seconds ... (USA).

12/04/1965 ... Gemini 7 ... Frank Borman, James Lovell ... This was the longest standing U.S. flight for eight years (206 orbits), a record up until the Soyuz 9 flight; rendezvous with Gemini 6. Duration was 13 days, 18 hours, 35 minutes ... (USA).

12/15/1965 ... Gemini 6 ... Walter Schirra, Thomas Stafford ... The first manned rendezvous, within two feet of Gemini 7 (the planned Agena package was lost). Duration was 1 day, 01 hour, 51 minutes ... (USA).

03/16/1966 ... Gemini 8 ... Neil Armstrong, David Scott ... The first docking (with an Agena package) of one space vehicle with another; emergency reentry after a control malfunction; first landing in the Pacific Ocean. Duration was 10 hours, 41 minutes ... (USA).

06/03/1966 ... Gemini 9 ... Thomas Stafford, Eugene Cernan ... 127 minute EVA by Cernan, but no docking with the target; the landing in the Atlantic Ocean was off by a half of a mile from the recovery ship. Duration was 3 days, 21 minutes ... (USA).

07/18/1966 ... Gemini 10 ... John Young, Michaels Collins ... Docked with Agena 10 and used its engine to achieve a record 474-mile (763-kilometer) altitude; rendezvous with Agena 8; 39-minute EVA and 49-minute SEVA by Collins. Duration was 2 days, 22 hours, 37 minutes ... (USA).

09/12/1966 ... Gemini 11 ... Charles Conrad, Richard Gordon ... Used the docked Agena engine to attain a new altitude record of 850 miles (1369 kilometer) in altitude; 33 minute EVA and 128 minute EVA by Gordon; connected Gemini and Agena by a tether to study the effects; first automatic computer-guided reentry. Duration was 2 days, 23 hours, 17 minutes, 9 seconds ... (USA).

11/01/1966 ... X-15 Flight 174 ... William H. Dana ... A sub-orbital flight of 58.1 miles (93.5 kilometers), and a speed of 3,750 mph (6,040 km/h) ... Duration was 10 minutes, 43.8 seconds ... (USA).

11/11/1966 ... Gemini 12 ... James Lovell, Edwin Aldrin ... The final Gemini mission; Agena docking; 126 minute EVA and two EVAs totaling

204 minutes by Aldrin; first work carried out during an EVA; automatic computer-guided reentry. Duration was 3 days, 22 hours, 35 minutes … (USA).

04/23/1967 … Soyuz 1 … Vladimir Komarov … The first reported flight of the Soyuz spacecraft system; Komarov died when the lines to the re-entry parachute became entangled, causing the capsule to hit the ground at around 200 mph, whereupon it was destroyed and burst into flames. Duration was 1 day, 02 hours, 48 minutes … (USSR).

10/17/1967 … X-15 Flight 190 … William "Pete" Knight ... A sub-orbital flight of 53.1 miles (85.5 kilometers), and a speed of 3,856 mph (6,206 km/h) … Duration was 10 minutes, 6.4 seconds … (USA).

11/17/1967 … X-15 Flight 191 … Michael J. Adams ... A sub-orbital flight of 50.3 miles (81.0 kilometers), and a speed of 3,569 mph (5744 km/h); a loss of flight control systems occurred during the reentry phase; the vehicle entered a spin at Mach 5 (five times the speed of sound), and at approximately 18,600 feet the vehicle began to dive followed by high frequency oscillations; the vehicle disintegrated when the forces reached 15Gs (15 times Earth's gravity), killing the astronaut … Duration was 4 minutes, 50.1 seconds … (USA).

03/02/1968 … Zond-4 L1 No61 spacecraft … Yuri Gagarin and Vladimir Seregin ... This was intended as a lunar trajectory mission that had onset failures from the start. The Soviet Union intended to show off their superiority by having the "first" human in space also become the first person to circle the Moon. Earth orbit was achieved, but an escape velocity to reach the Moon was impossible due to multiple failures of the engine systems. It became clear on reaching orbit that it was impossible to execute the flight's schedule or to return the cosmonauts to Earth safely. It was decided to suppress the cosmonauts' presence on board the spaceship in accordance with the practice to suppress the failures which then occurred from the Soviet Union's efforts. They flew on in nearly a complete radio-silence mode, talking with Earth by ultra-short wave transmissions only. Seven days after the launch, the spacecraft entered the dense layers of the upper atmosphere and burned up over the Bay of Guinea. Both cosmonauts perished.

What may have remained of the capsule and the two cosmonauts is presumed to be at the bottom of the Solomon Sea at the westernmost edge of the Pacific Ocean. To this date, there have been no recovery efforts made by the Russians or anyone else. The total duration for the flight has never been revealed, but believed to be (approximately) 7 days, 11 hours, 51 minutes ... (USSR).

08/21/1968 ... X-15 Flight 197 ... William H. Dana ... A sub-orbital flight of 50.6 miles (81.4 kilometers), and a speed of 3,443 mph (5541 km/h) ... Duration was 9 minutes, 23.3 seconds ...(USA).

10/11/1968 ... Apollo 7 ... Walter Schirra, Donn Eisele, Walter Cunningham ... The first piloted flight of the Apollo spacecraft, command and service module only; the first American three-man crew; the first live television footage on board Apollo. Duration was 10 days, 20 hours, 09 minutes ... (USA).

10/25/1968 ... Soyuz 2 ... Ivan Istochnikov ... denied by the Russians, who claimed that Soyuz 2 was an unmanned spacecraft, later reports indicated that it was indeed manned by Istochnokov and a dog, and that they died during the orbital flight out of the Baikonur space center; it was believed that, once again, there were some design failures with the parachute landing system. Duration was (approximately) 1 day, 16 hours, 33 minutes ... (USSR).

10/26/1968 ... Soyuz 3 ... Georgi Beregovoi ... There was an attempted rendezvous with Ivan Istochnikov on board Soyuz 2; it was only partially successful, with the two spacecraft never coming closer than 200 meters (approximately); having used up so much fuel during his attempts to perform an actual docking with Soyuz 2, Beregovoi had to give up the attempts. Duration was 3 days, 22 hours, 51 minutes ... (USSR).

12/21/1968 ... Apollo 8 ... Frank Borman, James Lovell, William Anders ... The first successful lunar orbit by the United States, and piloted lunar return reentry (the first attempts to orbit the Moon were made by the Russians using modified Vostok spacecraft: on 02/17/1961 Gennady Mikhailov made an attempt, and on 10/14/1961 Ivan Grachov made the second attempt; both men died; the Russians have never admitted these failures); the first manned Saturn V launch vehicle; the first televised views of the lunar surface. Duration was 6 days, 03 hours, 01 minutes … (USA).

01/14/1969 … Soyuz 4 … Vladimir Shatalov ... First docking of two piloted spacecraft; docked with Soyuz 5; transferred from Soyuz 5 onto Soyuz 4, were Aleksei Yeliseyev and Yevgeni Khrunov, who reentered and landed with Shatalov; this was a demonstration of technology some saw as a publicity stunt contrived to convince the world of Russia's technological superiority. Duration was 2 days, 23 hours, 21 minutes … (USSR).

01/15/1969 … Soyuz 5 … Boris Volynov, Aleksei Yeliseyev, Yevgeni Khrunov ... Docked with Soyuz 4, and Yeliseyev and Khrunov transferred to Soyuz 4 by way of an EVA for the reentry; Volynov landed by himself. Duration was 3 days, 54 minutes … (USSR).

03/03/1969 … Apollo 9 (CM - Gumdrop, LEM - Spider) … James McDivitt, David Scott, Russell Schweickart ... First piloted flight of the lunar module (in Earth orbit only); 46 minute EVA by Schweickart, who tested the lunar spacesuit; 61 minute EVA by Scott. Duration was 10 days, 01 hour, 01 minute … (USA).

05/18/1969 … Apollo 10 (CM - Charlie, LEM - Snoopy) … Thomas Stafford, John Young, Eugene Cernan ... First lunar module orbit of the Moon, with a descent down to 50,000 feet (15,000 meters) of the Moon's surface; new manned speed record of 6.8863 miles per second (11.0825 kilometers per second) at Earth's atmosphere reentry. Duration was 8 days, 03 minutes … (USA).

07/16/1969 … Apollo 11 (CM - Columbia, LEM - Eagle) … Neil Armstrong, Edwin Aldrin, Michael Collins ... First lunar landing, made by Armstrong and Aldrin (July the 20th, 1969); 151 minute lunar EVA; collected 48.5 pounds of soil and rock samples (22 kilograms); lunar stay time was 21 hours, 36 minutes. Duration was 8 days, 03 hours, 19 minutes … In an interesting side note to this mission, historian William Federer (of AmericanMinute.com), stated the following: "Before they got out of the Lunar Module, they had a moment of silence and Buzz Aldrin celebrates communion. He pours the grape juice in one-sixth gravity, and it does a slow little circle. He reads John 15: 'I am the vine, you are the branches.'

And he takes bread that was partly consumed at the communion before he launched … he saved a piece of the bread. And he celebrated communion. So the first items that were consumed on the Moon was communion." Of course, the world didn't hear about this, because a famous atheist had given NASA grief over Apollo 8 astronauts publicly reading from the Bible. Federer continued, "He asks for radio silence because Madalyn Murray O'Hair had threatened to sue because the Apollo 8 had mentioned God. They read from the book of Genesis. Lunar Module pilot William Anders began their Christmas Eve, 1968 broadcast, saying, "The crew of Apollo 8 has a message that we would like to send to you: 'In the beginning, God created the Heaven and the Earth. And the Earth was without form and void, and darkness was upon the face of the deep. And the spirit of God moved upon the face of the waters. And God said, 'Let there be light.' And there was light.'" And, as Apollo 11 headed back to Earth, Aldrin said to a listening world, "This has been far more than three men on a mission to the Moon. Personally, and reflecting on the events of the past several days, a verse from Psalms comes to mind: 'When I consider the Heavens, the works of Thy fingers, and the Moon and the stars, which Thou has ordained, what is man that Thou art mindful of him?' Nine months later, when an accident occurred on board Apollo 13 that might have doomed its three astronauts to die in space, much of the world turned to God. It was a rare moment when a united mankind thanked God! One of the last Apollo flights featured Mission Control's Charles Duke getting to walk on the Moon. He spoke of it later. "Charles Duke is a great Christian," Federer noted, "He's an astronaut and he says, 'I used to think that (paraphrasing here) going to the Moon would be my greatest achievement. But my walk with Jesus is more memorable because it's an everyday affair.' "Just a fascinating faith that Charles Duke had," Federer continued. "Plus all the astronauts: Jim Irwin on Apollo 15 became an evangelical minister. And Apollo14 left a microfilm copy of the King James Bible on the Moon." It's interesting to note how many of those men who flew far into the Heavens couldn't get the God of Heaven out of their thought … (USA).

10/11/1969 … Soyuz 6 … Georgi Shonin, Valeri Kubasov … Rendezvoused with Soyuz 7 and Soyuz 8; first triple rendezvous, a publicity stunt more than anything; first welding of metals in space. Duration was 4 days, 22 hours, 43 minutes … (USSR).

10/12/1969 … Soyuz 7 … Anatoli Filipchenko, Vlasislav Volkov, Viktor Gorbatko ... Triple rendezvous with Soyuz 6 and Soyuz 8; space laboratory construction tests were conducted; first time three manned spacecraft, and seven crew members, orbited the Earth simultaneously. Duration was 4 days, 22 hours, 40 minutes … (USSR).

0/13/1969 … Soyuz 8 … Vladimir Shatalov, Aleksei Yeliseyev ... Triple rendezvous with Soyuz 6 and Soyuz 7; part of the space laboratory construction team. Duration was 4 days, 22 hours, 51 minutes … (USSR).

11/14/1969 … Apollo 12 (CM - Yankee Clipper, LEM - Intrepid) … Charles Conrad, Richard Gordon, Alan Bean ... The second Moon landing, with Conrad and Bean; two lunar EVAs totaling 465 minutes; collected 74.7 pounds of samples (33.9 kilograms); total time on the lunar surface was 31h:31m; total duration for the flight was 10 days, 04 hours, 36 minutes … (USA).

04/11/1970 … Apollo 13 (CM - Odyssey, LEM - Aquarius) … James Lovell, John Swigert, Fred Haise ... The mission was aborted following an explosion of an oxygen tank in the service module; crew returned to Earth using the lunar module for much of the flight and the command module for the reentry; circumlunar return; manned altitude record of 248,665 miles (400,187 kilometers) above the Earth's surface; this was a near-catastrophic flight. Duration was 5 days, 22 hours, 55 minutes … (USA).

06/01/1970 … Soyuz 9 … Adrian Nikolayev, Vitali Sevestyanov ... This flight marked the start of working under weightless conditions for the Russians; endurance record for a solo spacecraft. Duration was 17 days, 16 hours, 59 minutes … (USSR).

01/31/1971 … Apollo 14 (CM - Kitty Hawk, LEM - Antares) … Alan Shepard, Edgar Mitchell, Stuart Roosa ... Shepard and Mitchell made the third Moon landing; two lunar EVAs totaling 563 minutes; collected 96 pounds (43.5 kilograms) of lunar samples; lunar stay time was 33h:31m; total flight duration was 9 days, 02 minutes … (USA).

04/23/1971 … Soyuz 10 … Vladimir Shatalov, Aleksei Yeliseyev, Nikolai Rukavishnikov ... Adjustments made to an improved docking bay between the spacecraft and the orbiting Salyut space station, but the cosmonauts did not enter the station. Duration was 1 day, 23 hours, 46 minutes … (USSR).

06/06/1971 … Soyuz 11 … Georgi Dobrovolski (Dobrovolskiy; Dobrovolsky), Vladislav Volkov, Viktor Patsayev … Docked and entered the orbiting Salyut 1 space station; orbited inside Salyut for 23 days; the three crew members died on their return to Earth due primarily to the complete loss of pressurization, as the hatch seal was not properly aligned; the flight crew members were not wearing pressurized space suits during the reentry. Duration was 23 days, 18 hours, 22 minutes … (USSR).

07/26/1971 … Apollo 15 (CM - Endeavor, LEM - Falcon) … David Scott, James Irwin, Alfred Worden ... Scott and Irwin made the fourth landing on the Moon; first use of a lunar rover for transportation on the lunar surface; first deep spacewalk; three lunar EVAs totaling 19 hours, 8 minutes; EVA made by Scott; collected 170 pounds (77 kilograms) of samples; lunar stay time was 66 hours, 54 minutes; Worden made a 38 minute EVA; Irwin made a EVA; a sub satellite was released. Duration was 12 days, 07 hours, 12 minutes … (USA).

04/16/1972 … Apollo 16 (CM - Casper, LEM - Orion) … John Young, Thomas Mattingly, Charles Duke ... Young and Duke made the fifth Moon landing; three lunar EVAs totaling 20 hours, 14 minutes; collected 213 pounds (97 kilograms) of lunar samples; lunar stay time was 71 hours, 14 minutes; Mattingly made a 73 minute EVA, while Duke made another EVA; sub satellite was released; Duration was 11 days, 01 hour, 51 minutes … (USA).

12/07/1972 … Apollo 17 (CM - America, LEM - Challenger) … Eugene Cernan, Ronald Evans, Harrison Schmitt ... Cernan and Schmitt made the sixth and last Apollo lunar landing; three lunar EVAs totaling 22 hours, 4 minutes; collected 243 pounds (110 kilograms) of samples; record lunar stay of 74 hours, 59 minutes; Evans made a 66 minute EVA, and Schmitt made another EVA. Duration was 12 days, 13 hours, 52 minutes … (USA).

05/25/1973 ... Apollo Skylab 2 ... Charles Conrad, Joseph Kerwin, Paul Weitz ... First American piloted space station; conducted long-term flight tests; crew repaired damage caused during the launch phase of the station; two EVAs totaling 5 hours, 14 minutes; one 75 minute SEVA. Duration was 28 days, 50 minutes ... (USA).

07/28/1973 ... Apollo Skylab 3 ... Alan Bean, Owen Garriott, Jack Lousma ... Crew systems and operational tests; exceeded pre-mission plans for scientific activities; three EVAs totaling 13 hours, 46 minutes. Duration was 59 days, 11 hours, 10 minutes ... (USA).

09/27/1973 ... Soyuz 12 ... Vasili Lazarev, Oleg Makarov ... New life support equipment was tested following the Soyuz 11 tragedy; full pressure suits were mandated to be worn. Duration was 1 day, 23 hours, 16 minutes ... (USSR).

11/16/1973 ... Apollo Skylab 4 ... Gerald Carr, Edward Gibson, William Pogue ... The final Skylab mission; duration record set until the Soyuz 26-Salyut 6 mission; four EVAs totaling 22 hours, 21 minutes; set the then-record spacewalk of 7 hours, 1 minute. Duration was 84 days, 01 hour, 15 minutes ... (USA).

12/18/1973 ... Soyuz 13 ... Pyotr Klimuk, Valentin Lebedev ... Astrophysical observations were made with the Orion ultraviolet telescope; sections of the Earth were photographed. Duration was 7 days, 20 hours, 56 minutes ... (USSR).

07/03/1974 ... Soyuz 14 ... Pavel Popovich, Yuri Artyukhim ... The sole occupation of Salyut 3; no further dockings were made. Duration was 15 days, 17 hours, 30 minutes ... (USSR).

08/26/1974 ... Soyuz 15 ... Gennadi Sarafanov, Lev Demin ... Failed to dock with Salyut 3 because of systems failures on board of Soyuz 15; first return to Earth during the night. Duration was 2 days, 12 minutes ... (USSR).

11/05/1974 ... FSW ... Liu Chongfu ... This was to be China's second attempt at human space flight, and was only intended as a quick sub-

orbital mission, but the first stage of the rocket exploded seconds after the initial launch, which instantly killed the astronaut, as there was no escape system in place. In the same fashion as the Soviets, the Chinese have denied that these flights ever took place! The flight duration was approximately five seconds … (China).

12/02/1974 … Soyuz 16 … Anatoli Filipchenko, Nikolai Rukavishnikov … This was a trial run with a modified Soyuz capable of docking with the U.S. Apollo spacecraft. Duration was 5 days, 22 hours, 24 minutes … (USSR).

01/11/1975 … Soyuz 17 … Aleksey Gubarev, Georgi Grechko … First occupation of Salyut 4; Soviet endurance record. Duration was 29 days, 13 hours, 20 minutes … (USSR).

04/05/1975 … Soyuz 18-A … Vasili Lazarev, Oleg Makarov … The launch towards Salyut 4 was aborted when the first stage of the booster rocket failed to separate; there was an immediate emergency reentry which caused 14g forces on the two cosmonauts; due to the emergency, this in essence turned out to be a sub-orbital flight. Duration was 21 minutes, 27 seconds … (USSR).

05/24/1975 … Soyuz 18-B … Pyotr Klimuk, Vitali Sevestyanov … The second Salyut 4 occupation; a new Soviet endurance record was set. Duration was 62 days, 23 hours, 20 minutes … (USSR).

07/15/1975 … Apollo 18 (ASTP) … Thomas Stafford, Vance Brand, Donald Slayton … Joint flight and docking with Soyuz 19; docked with the Soyuz vessel for two days; shared meals and held a joint news conference. Duration was 9 days, 01 hour, 28 minutes … (USA).

07/15/1975 … Soyuz 19 (ASTP) … Alexsei Leonov, Valeri Kubasov … Joint flight and docking with Apollo 18. Duration was 5 days, 22 hours, 21 minutes … (On 11/17/1975 … Soyuz 20 … (unmanned)). Docked with the Salyut 4 station and used as a long duration test platform; contained living organisms; it was recovered from orbit on February the 16th, 1976) … (USSR).

11/26/1975 … FSW … Yang Xiaohai … After the death of their second astronaut a year earlier, they had a successful orbital flight, with what was believed to be three orbits, but the tiny spacecraft crashed to the Earth in Mongolia when the parachute lines became tangled, which in turn caused the spacecraft to hit the ground at nearly 200 miles per hour, instantly killing the astronaut. And there was nothing in the way of retro-rockets to slow the descent, so it was a recipe for disaster from the very beginning. The Chinese stopped their attempts for human spaceflight after this failure, and waited for nearly 28 years before any other attempts were made. The duration was (approximately) 04 hours, 36 minutes … (China).

07/06/1976 … Soyuz 21 … Boris Volynov, Vitali Zholobov ... First Salyut 5 occupation; acid fumes on board the station forced the evacuation and return to Earth of the cosmonauts. Duration was 49 days, 06 hours, 24 minutes … (USSR).

09/15/1976 … Soyuz 22 … Valeri Bykovsky, Vladimir Aksyonov ... This was a slightly modified Soyuz capsule in order that the Earth's surface could be photographed from orbit; the flight was cut short because of the crew's inability to rendezvous with Salyut 5. Duration was 7 days, 21 hours, 52 minutes … (USSR).

10/14/1976 … Soyuz 23 … Vyacheslav Zudov, Valeri Rozhdestvensky ... Docking between the spacecraft and the Salyut 5 failed; this was the first Soviet splashdown after an emergency return to Earth, landing on the partially frozen Lake Tengiz; recovery took over nine hours due to adverse weather conditions at the landing site. Duration was 2 days, 07 minutes … (USSR).

02/07/1977 … Soyuz 24 … Victor Gorbatko, Yuri Glazkov ... This was the second crew to occupy the Salyut 5 space station. Duration in space was 17 days, 17 hours, 26 minutes … (USSR).

10/09/1977 … Soyuz 25 … Vladimir Kovalyonok, Valeri Ryumin ... This was a failure, as the planned manual docking with the orbiting Salyut 5 could not be accomplished. Duration was 2 days, 45 minutes … (USSR).

12/10/1977 ... Soyuz 26 ... Yuri Romanenko, Georgi Grechko ... The first crew occupation of the Salyut 6 station after a successful docking; Romanenko and Grechko would return to Earth on board Soyuz 27; Progress 1 was an unmanned vessel that would resupply the Salyut. Duration was 37 days, 10 hours, 06 minutes ... (USSR).

01/10/1978 ... Soyuz 27 ... Vladimir Dzhanibekov, Oleg Makarov ... The second crew occupancy for Salyut 6; dual crew occupancy of the station; Dzhanibekov and Makarov would return to Earth on board Soyuz 26. Duration was 64 days, 22 hours, 53 minutes ... (USSR).

03/02/1978 ... Soyuz 28 ... Aleksey Gubarev, Vladimir Remek ... First international crew (Russian and Czechoslovakian) to dock with the orbiting Salyut 6 station. Duration was 7 days, 22 hours, 16 minutes ... (USSR).

06/15/1978 ... Soyuz 29 ... Vladimir Kovalynok, Aleksandr Ivanchenkov ... First 100+ day flight to the Salyut 6 station; Kovalynok and Ivanchenkov would return to Earth on board Soyuz 31; Progress 2, 3 and 4 would resupply the station. Duration for Soyuz 29 was 79 days, 15 hours, 24 minutes ... (USSR).

06/27/1978 ... Soyuz 30 ... Pyotr Klimuk, Miroslav Hermaszewski ... Second international crew (Russian and Polish) to the Salyut 6 station. Duration was 7 days, 22 hours, 03 minutes ... (USSR).

08/26/1978 ... Soyuz 31 ... Valeri Bykovsky, Sigmund Jaehn ... Third international crew (Russian and East German) to the Salyut 6 station; Bykovsky and Jaehn would return to Earth on board Soyuz 29. Duration was 67 days, 20 hours, 13 minutes ... (USSR).

02/25/1979 ... Soyuz 32 ... Vladimir Lyakhov, Valeri Ryumin ... They docked with Salyut 6; endurance record would be set; they would return to Earth on board Soyuz 34; Progress 5, 6 and 7 would resupply the Salyut while they were there. Duration was 108 days, 04 hours, 25 minutes ... (USSR).

04/10/1979 ... Soyuz 33 ... Nikolai Rukavishnikov, Georgi Ivanov ... The fourth international crew (Russian and Bulgarian); failed to dock with

Salyut 6 due to an engine malfunction. Duration was 1 day, 23 hours, 01 minute … (USSR).

06/06/1979 … Soyuz 34 … unmanned vessel launched to Salyut 6 in order for two cosmonauts (Vladimir Lyakhov and Valeri Ryumin) to have a return-to-Earth vessel; compensated for the failure of Soyuz 33. Duration was 74 days, 18 hours, 17 minutes … (USSR).

12/16/1979 … Soyuz T-1 … an unmanned test of a modified Soyuz vessel; the first attempt to dock with Salyut failed as it overshot its target, but a second attempt three days later was successful; it used its engines to raise the orbital height of the Salyut; the 'T' stood for 'Transport'; first use of extended solar panels for electricity on a Soyuz; it was sent back to Earth on March the 25th, 1980 … (USSR).

04/09/1980 … Soyuz 35 … Leonid Popov, Valeri Ryumin ... The fourth long-duration stay on board Salyut 6; Popov and Ryumin would return on board Soyuz 37; the Salyut was resupplied by Progress 8, 9 and 11 while they were on board. Duration was 55 days, 01 hour, 28 minutes … (USSR).

05/26/1980 … Soyuz 36 … Valeri Kubasov, Bertalan Farkas ... The fifth international crew (Russian and Hungarian); docked with Salyut 6; Kubasov and Farkas would return to Earth on board Soyuz 35. Duration was 65 days, 20 hours, 54 minutes … (USSR).

06/05/1980 … Soyuz T-2 … Yuri Malyshev, Vladimir Aksyonov ... The first manned flight of the modified Soyuz spacecraft; manual docking with Salyut 6. Duration was 3 days, 22 hours, 20 minutes … (USSR).

07/23/1980 … Soyuz 37 … Victor Gorbatko, Pham Tuan ... The sixth international crew (Russian and Vietnamese); docked with Salyut 6; Gorbatko and Tuan would return to Earth on board Soyuz 36. Duration was 79 days, 15 hours, 17 minutes … (USSR).

09/18/1980 …Soyuz 38 … Yuri Romanenko, Arnaldo Tamayo-Mendez ... The seventh international crew (Russian and Cuban); docked with the orbiting Salyut 6 complex. Duration was 7 days, 20 hours, 43 minutes … (USSR).

11/27/1980 … Soyuz T-3 … Leonid Kizim, Oleg Makarov, Gennadi Strekalov … The resumption of three-man flights; repair work accomplished to Salyut 6; Progress 11 resupplied Salyut 6 while this crew was on board. Duration was 12 days, 19 hours, 08 minutes … (USSR).

03/12/1981 … Soyuz T-4 … Vladimir Kovalyonok, Viktor Savinykh … The last long duration stay on board Salyut 6; Progress 12 resupplied the Salyut while the crew was there. Duration was 74 days, 17 hours, 37 minutes … (USSR).

03/22/1981 … Soyuz 39 … Vladimir Dzhanibekov, Jugderdemidiyn Gurragcha … The eighth international crew (Russian and Mongolian); docked with Salyut 6. Duration was 7 days, 20 hours, 42 minutes … (USSR).

04/12/1981 … STS-1 Columbia … John Young, Robert Crippen … The first flight of the U.S. shuttle spacecraft system. Duration was 2 days, 06 hours, 22 minutes … (USA).

05/14/1981 … Soyuz 40 … Leonid Popov, Dumitru Prunariu … The ninth international crew (Russian and Romanian); docked with Salyut 6; this was the last flight of the old Soyuz design. Duration was 7 days, 21 hours, 42 minutes … (USSR).

11/12/1981 … STS-2 Columbia … Joseph Engle, Richard Truly … The first reuse of the shuttle system, and the second orbital test flight; first test of the Canadian robotic arm (RMS); the five day mission was cut short due to a fuel cell malfunction. Duration was 2 days, 06 hours, 14 minutes … (USA).

03/22/1982 … STS-3 Columbia … Jack Lousma, Gordon Fullerton … Third orbital test flight of the shuttle; first experiments; astronomical payload; the landing was delayed for one day due to a storm. Duration was 8 days, 06 minutes … (USA).

05/13/1982 … Soyuz T-5 … Anatoli Berezovoi, Valentin Lebedev … The first Salyut 7 occupation; the crew would return to Earth on board

Soyuz T-7; Progress 13, 14, 15 and 16 supplied the Salyut 7 during this time. Duration was 106 days, 05 hours, 06 minutes … (USSR).

06/24/1982 … Soyuz T-6 … Vladimir Dzhanibekov, Aleksandr Ivanchenkov, Jean-Loup Chretien ... The tenth international crew (Russian and French); docked with Salyut 7. Duration was 7 days, 21 hours, 51 minutes … (USSR).

06/27/1982 … STS-4 Columbia … Thomas Mattingly, Henry Hartsfield ... The last orbital test flight for the shuttle system; first landing on a concrete runway; SRBs were lost; military payload; the beginning of operational flights. Duration was 7 days, 01 hour, 11 minutes … (USA).

08/19/1982 … Soyuz T-7 … Leonid Popov, Alexsandr Serebrov, Svetlana Savitskaya ... Docked with Salyut 7; Savitskaya would be the first female on board a Soyuz and a Salyut system; the crew returned to Earth on board Soyuz T-5. Duration was 113 days, 01 hour, 51 minutes … (USSR).

11/11/1982 … STS-5 Columbia … Vance Brand, Robert Overmyer, Joseph Allen, William Lenoir ... The first operational shuttle mission; the first four-man crew; deployed two communications satellites (COMSATs); EVA performed. Duration was 5 days, 02 hours, 15 minutes … (USA).

04/04/1983 … STS-6 Challenger … Paul Weitz, Karol Bobko, Donald Peterson, Story Musgrave ... First flight of the Challenger shuttle; first shuttle EVA; released a Tracking and Data Recovery Satellite (TDRS). Duration was 5 days, 25 minutes … (USA).

04/20/1983 … Soyuz T-8 … Vladimir Titov, Gennadi Strekalov, Aleksandr Serebrov ... Failed to dock with Salyut 7; complete failure of the radar; manual approach to the orbiting station was aborted. Duration was 2 days, 18 minutes … (USSR).

06/18/1983 … STS-7 Challenger … Robert Crippen, Frederick Hauck, Sally Ride, John Fabian, Norman Thagard ... Sally Ride was the first American woman in space; first five-person crew for the shuttle; two COMSAT satellites were released; German platform SPAS-1. Duration was 6 days, 02 hours, 25 minutes … (USA).

06/27/1983 ... Soyuz T-9 ... Vladimir Lyakhov, Aleksandr Paviovich Alexandrov ... Docked with Salyut 7; two EVAs, which added two solar panels; Progress 18 resupplied the orbiting Salyut 7. Duration was 149 days, 10 hours, 46 minutes ... (USSR).

08/30/1983 ... STS-8 Challenger ... Richard Truly, Daniel Brandenstein, Dale Gardner, Guion Bluford, William Thornton ... The first nighttime launch; Bluford was the first black astronaut in space; one communications satellite was deployed. Duration was 6 days, 01 hour, 10 minutes ... (USA).

09/27/1983 ... Soyuz T-10-A ... Vladimir Titov, Gennadi Strekalov ... A complete systems failure; launch pad fire and the first use of the escape tower; the descent module would be re-used by Soyuz T-15. Duration was 05 minutes, 13 seconds ... (USSR).

11/28/1983 ... STS-9/41-A Columbia ... John Young, Brewster Shaw, Owen Garriott, Robert Parker, Byron Lichtenberg, Ulf Merbold ... First six-person crew; Merbold was the first German on a U.S. mission; first Spacelab (SL-1) mission. Duration was 10 days, 07 hours, 48 minutes ... (USA).

02/03/1984 ... STS-41-B Challenger ... Vance Brand, Robert Gibson, Bruce McCandless, Ronald McNair, Robert Stewart ... The first untethered EVA and testing of the MMU jetpack; first landing at the Kennedy Space Center in Florida; two COMSAT satellites were released. Duration was 7 days, 23 hours, 17 minutes ... (USA).

02/08/1984 ... Soyuz T-10B ... Leonid Kizim, Vladimir Solovyov, Oleg Atkov ... The first long-stay triple crew to Salyut 7; performed six EVAs totaling 22 hours, 56 minutes; the three-man crew returned to Earth on board Soyuz T-11; Duration was 62 days, 22 hours, 41 minutes ... (USSR).

04/03/1984 ... Soyuz T-11 ... Yuri Malyshev, Gennadi Strekalov, Rakesh Sharma ... The eleventh international crew (Russian and Indian); docked with Salyut 7; the crew returned to Earth on board Soyuz T-10B. Duration was 181 days, 21 hours, 49 minutes ... (USSR).

04/06/1984 ... STS-41-C Challenger ... Robert Crippen, Francis Scobee, James van Hoften, Terry Hart, George Nelson ... First in-orbit retrieval and repair of a satellite (SMM); LDEF deployment. Duration was 6 days, 23 hours, 41 minutes ... (USA).

07/17/1984 ... Soyuz T-12 ... Vladimir Dzhanibekov, Svetlana Savitskaya, Igor Volk ... Docked with Salyut 7; first EVA by a female (Svetlana Savitskaya); EVA was for three hours and 45 minutes. Duration 11 days, 19 hours, 15 minutes ... (USSR).

08/30/1984 ... STS-41-D Discovery ... Henry Hartsfield, Michael Coats, Judith Resnik, Steven Hawley, Richard Mullane, Charles Walker ... The first flight of the Discovery shuttle; first commercial payload specialist on board; three communications satellites were deployed. Duration was 6 days, 57 minutes ... (USA).

10/05/1984 ... STS-41-G Challenger ... Robert Crippen, Jon McBride, Sally Ride, Kathryn Sullivan, David Leestma, Marc Garneau, Paul Scully-Power ... Garneau was the first Canadian astronaut in space; first seven-person crew; first U.S. female EVA by Sullivan; first two woman flight; environmental satellite deployed (ERBS). Duration was 8 days, 05 hours, 25 minutes ... (USA).

11/08/1984 ... STS-51-A Discovery ... Frederick Hauck, David Walker, Dale Gardner, Joseph Allen, Anna Fisher ... First satellite retrieval and return to Earth for study; two COMSAT satellites deployed. Duration was 7 days, 23 hours, 46 minutes ... (USA).

01/24/1985 ... STS-51-C Discovery ... Thomas Mattingly, Loren Shriver, Ellison Onizuka, James Buchli, Gary Payton ... First mission dedicated to the Department of Defense (DOD); classified military payload was deployed. Duration was 3 days, 01 hour, 34 minutes ... (USA).

04/12/1985 ... STS-51-D Discovery ... Karol Bobko, Donald Williams, Rhea Seddon, Jeffery Hoffman, David Griggs, Charles Walker, Jake Garn ... Garn was the first U.S. Senator in space; two COMSAT satellites were deployed; an EVA was accomplished to repair the Syncom IV satellite. Duration was 6 days, 23 hours, 55 minutes, 23 seconds ... (USA).

04/29/1985 ...STS-51-B Challenger ... Robert Overmyer, Frederick Gregory, Don Lind, Norman Thagard, William Thornton, Lodewijk van den Berg, Taylor Wang ... The second Spacelab science mission; one satellite was deployed. Duration was 7 days, 10 minutes ... (USA).

06/06/1985 ... Soyuz T-13 ... Vladimir Dzhanibekov, Viktor Savinykh ... Reactivation of the Salyut 7 orbiting laboratory; Savinykh stayed with the Soyuz T-14 crew. Duration was 112 days, 03 hours, 12 minutes ... (USSR).

06/17/1985 ... STS-51-G Discovery ... Daniel Brandenstein, John Creighton, Shannon Lucid, John Fabian, Steven Nagel, Patrick Baudry, Sultan Salman Al-Saud ... Salman Al-Saud was the first Arab in space; Baudry was the first Frenchman on a U.S. mission; three communication satellites were deployed. Duration was 7 days, 01 hour, 40 minutes ... (USA).

07/29/1985 ... STS-51-F Challenger ... Gordon Fullerton, Roy Bridges, Story Musgrave, Loren Acton, Anthony England, Karl Henize, David Bartoe ... The third Spacelab astronomy and science mission. Duration was 7 days, 22 hours, 46 minutes ... (USA).

08/27/1985 ... STS-51-I Discovery ... Joseph Engle, Richard Covey, James van Hoften, John Lounge, William Fisher ... Three COMSAT satellites were deployed; an EVA was performed to repair the Syncom IV satellite already in orbit. Duration was 7 days, 02 hours, 18 minutes ... (USA).

09/17/1985 ... Soyuz T-14 ... Vladimir Vasyutin, Georgi Grechko, Aleksandr Volkov ... Docked with the Salyut 7; first mission to be aborted due to an illness (Vasyutin); Grechko returned with Soyuz T-13. Duration was 64 days, 21 hours, 52 minutes ... (USSR).

10/03/1985 ... STS-51-J Atlantis ... Karol Bobko, Ronald Grabe, David Hilmers, Robert Stewart, William Pailes ... The first flight of the Atlantis shuttle; the second Department of Defense (DOD) mission; details of the flight remain classified. Duration was 4 days, 01 hour, 46 minutes ... (USA).

10/30/1985 ... STS-61-A Challenger ... Henry Hartsfield, Steven Nagel, James Buchli, Guion Bluford, Bonnie Dunbar, Reinhard Furrer, Ernst Messerschmidt, Wubbo Ockels ... The first eight-person crew; the first German Spacelab mission (D1). Duration was 7 days, 46 minutes ... (USA).

11/26/1985 ... STS-61-B Atlantis ... Brewster Shaw, Bryan O'Connor, Mary Cleave, Sherwood Spring, Jerry Ross, Rodolfo Neri-Vela, Charles Walker ... Neri-Vela was the first Mexican in space; three COMSAT satellites were deployed; space structures assembly tests were conducted. Duration was 6 days, 21 hours, 06 minutes ... (USA).

01/12/1986 ... STS-61-C Columbia ... Robert Gibson, Charles Bolden, Steven Hawley, Franklin Chang-Diaz, George Nelson, Robert Cenker, William Nelson ... Bill Nelson was the first U.S. Congressman in space; one COMSAT satellite was deployed; material and astronomy experiments were conducted. Duration was 6 days, 02 hours, 05 minutes ... (USA).

01/28/1986 ... STS-51-L Challenger ... Francis Scobee, Michael Smith, Judith Resnik, Ellison Onizuka, Ronald McNair, Gregory Jarvis, Christie McAuliffe ... The shuttle blew up shortly after launch, killing all seven astronauts; the cause of the tragedy was the apparent failure of a rubberized collar on one of the booster rockets, due to the freezing weather conditions; the flames from the booster rocket ignited the solid propellant tank upon which the shuttle was riding, causing the explosion. Duration was 01 minute, 13 seconds ... (USA).

02/20/1986 ... Mir ... a new generation unmanned space station with six docking ports was launched; it would be occupied by multiple crews. Total duration in orbit was 5,510 days ... (USSR).

03/13/1986 ... Soyuz T-15 ... Leonid Kizim, Vladimir Solovyov ... The first crew to enter the Mir station; also made a successful excursion to Salyut 7 from May the 5th to June the 26th; two Salyut EVAs totaling 8 hours, 50 minutes; the crew re-used the descent module from the previous T-10A abort. Duration was 125 days, 01 minute ... (USSR).

05/21/1986 ... Soyuz TM-1 ... an unmanned test flight of the Soyuz TM vessel, intended as a replacement for the Soyuz T series used in the Salyut missions ... (USSR).

02/06/1987 ... Soyuz TM-2 ... Yuri Romanenko, Aleksandr Laviekin ... The second Mir crew occupation; Romanenko remained at the Mir station and set a new endurance record; he returned to Earth on board Soyuz TM-3. Duration was 174 days, 03 hours, 26 minutes ... (USSR).

07/22/1987 ... Soyuz TM-3 ... Aleksandr Viktorenko, Aleksandr Pavlovich Alexandrov, Mohammed Faris ... Docked with the Mir station; Aleksandrov remained in the station, replacing Laveykin. Duration was 160 days, 07 hours, 17 minutes ... (USSR).

12/21/1987 ... Soyuz TM-4 ... Vladimir Titov, Musa Manarov, Anatoli Levchenko ... The third Mir long-stay mission; Titov and Manarov swapped with Romanenko and Aleksandrov; endurance record set; three EVAs totaling thirteen hours, forty minutes; returned in Soyuz TM-6. Duration was 178 days, 22 hours, 54 minutes ... (USSR).

06/07/1988 ... Soyuz TM-5 ... Anatoliy Solovyov, Viktor Savinykh, Aleksandr Panayotov Alexandrov ... Docked with Mir; the crew returned to Earth on board Soyuz TM-4. Duration was 91 days, 10 hours, 46 minutes ... (USSR).

08/29/1988 ... Soyuz TM-6 ... Vladimir Lyakhov, Valeri Polyakov, Abdul Mohmand ... Docked with Mir; the fourth long-stay mission; Polyakov stayed with Titov and Manarov. Duration was 114 days, 05 hours, 34 minutes ... (USSR).

09/29/1988 ... STS-26 Discovery ... Frederick Hauck, Richard Covey, David Hilmers, John Lounge, George Nelson ... The redesigned shuttle makes its first flight; deployed the Tracking Data Relay Satellite (TDRS-C). Duration was 4 days, 01 hour, 01 minute ... (USA).

11/26/1988 ... Soyuz TM-7 ... Aleksandr Volkov, Sergei Krikalyov, Jean-Loup Chretien ... The fifth Mir long-stay mission; Volkov and Krikalyov swapped with Titov and Manarov; Chretien performed an EVA, and returned to Earth in Soyuz TM-6; Volkov and Krikalyov returned to Earth in Soyuz TM-7. Duration was 151 days, 11 hours, 08 minutes … (USSR).

12/02/1988 … STS-27 Atlantis … Robert Gibson, Guy Gardner, Richard Mullane, William Shepherd, Jerry Ross … The third Department of Defense (DOD) classified mission; the first Lacrosse imaging radar satellite for all-weather day and night reconnaissance was deployed. Duration was 4 days, 09 hours, 06 minutes … (USA).

03/13/1989 … STS-29 Discovery … Michael Coats, John Blaha, James Buchli, Robert Springer, James Bagian … Deployed the TDRS-D Data Relay Satellite; conducted scientific experiments. Duration was 4 days, 23 hours, 40 minutes … (USA).

05/04/1989 … STS-30 Atlantis … David Walker, Ronald Grabe, Norman Thagard, Mary Cleaver, Mark Lee … Deployed the Magellan Venus orbiter (launched on the IUS stage); the Magellan arrived at Venus in August of 1999. Duration was 4 days, 58 minutes … (USA).

08/08/1989 … STS-28 Columbia … Brewster Shaw, Richard Richards, James Adamson, David Leestma, Mark Brown … The fourth Department of Defense (DOD) classified mission; deployed a military satellite for relaying reconnaissance imagery. Duration was 5 days, 01 hour, 01 minute … (USA).

09/05/1989 … Soyuz TM-8 … Aleksandr Viktorenko, Aleksandr Serebrov … The fifth long-stay mission on board the Mir; received the Kvant 2 module; performed five EVAs, including two with maneuvering backpacks. Duration was 166 days, 06 hours, 58 minutes … (USSR).

10/18/1989 … STS-34 Atlantis … Donald Williams, Michael McCulley, Franklin Chang-Diaz, Shannon Lucid, Ellen Baker … The Galileo Jupiter orbiter was deployed from the IUS stage; Galileo arrived at the planet Jupiter in December of 1995. Duration was 4 days, 23 hours, 40 minutes … (USA).

11/22/1989 ... STS-33 Discovery ... Frederick Gregory, John Blaha, Story Musgrave, Manley Carter, Kathryn Thornton ... The fifth Department of Defense (DOD) classified mission; deployed at least one intelligence gathering satellite. Duration was 5 days, 08 minutes ... (USA).

01/09/1990 ... STS-32 Columbia ... Daniel Brandenstein, James Wetherbee, Bonnie Dunbar, Marsha Ivins, David Low ... Deployed one COMSAT satellite; retrieved from orbit and returned to Earth the Long Duration Exposure Facility (LDEF). Duration was 10 days, 21 hours, 02 minutes ... (USA).

02/11/1990 ... Soyuz TM-9 ... Anatoliy Solovyov, Aleksandr Balandin ... The sixth Mir long-stay mission; received the Kristall unmanned package for the station; two EVAs were performed. Duration was 179 days, 01 hour, 18 minutes ... (USSR).

02/28/1990 ... STS-36 Atlantis ... John Creighton, John Casper, Pierre Thuot, Richard Mullane, David Hilmers ... The sixth Department of Defense (DOD) classified mission; deployed a reconnaissance satellite. Duration was 4 days, 10 hours, 19 minutes ... (USA).

04/24/1990 ... STS-31 Discovery ... Loren Shriver, Charles Bolden, Steven Hawley, Kathryn Sullivan, Bruce McCandless ... Deployed the Hubble Space Telescope (HST); set an altitude record for the shuttle at 385 miles (619 kilometers. Duration was 5 days, 01 hour, 17 minutes ... (USA).

08/01/1990 ... Soyuz TM-10 ... Gennadi Manakov, Gennadi Strekalov ... The seventh long-stay Mir mission; emphasized Kristall materials processing; one EVA was performed. Duration was 130 days, 20 hours, 36 minutes ... (USSR).

10/06/1990 ... STS-41 Discovery ... Richard Richards, Robert Cabana, Bruce Melnick, William Shepherd, Thomas Akers ... The Ulysses solar probe was launched on the IUS stage. Duration was 4 days, 02 hours, 11 minutes ... (USA).

11/15/1990 ... STS-38 Atlantis ... Richard Covey, Frank Culbertson, Charles Gemar, Carl Meade, Robert Springer ... The seventh Department of

Defense (DOD) classified mission; deployed an intelligence gathering satellite. Duration was 4 days, 21 hours, 55 minutes … (USA).

12/02/1990 …STS-35 Columbia … Vance Brand, Guy Gardner, Jeffery Hoffman, John Lounge, Robert Parker, Samuel Durrance, Ronald Parise … Astronomy Spacelab ASTRO; UV/X-ray telescope. Duration was 8 days, 23 hours, 06 minutes … (USA).

12/02/1990 … Soyuz TM-11 … Viktor Afanasyev, Musa Manarov, Toyohiro Akiyama … The eighth Mir long-stay mission; four EVAs were performed; Akiyama of Japan returned to Earth on board Soyuz TM-10. Duration was 175 days, 01 hour, 51 minutes … (USSR).

04/05/1991 … STS-37 Atlantis … Steven Nagel, Kenneth Cameron, Linda Godwin, Jerry Ross, Jay Apt … Deployed the Gamma Ray Observatory (GRO); two EVAs were performed (the first since 1985). Duration was 5 days, 23 hours, 34 minutes … (USA).

04/28/1991 … STS-39 Discovery … Michael Coats, Blaine Hammond, Gregory Harbaugh, Donald McMonagle, Guion Bluford, Charles Veach, Richard Hieb … Infrared astronomy experiments were conducted. Duration was 8 days, 07 hours, 23 minutes … (USA).

05/18/1991 … Soyuz TM-12 … Anatoli Artsebarsky, Sergei Krikalyov, Helen Sharman … The ninth Mir long-stay mission; Sharman (UK) returned to Earth on board Soyuz TM-11; Artesbarsky returned on board Soyuz TM-12; Krikalyov returned on board Soyuz TM-13; six EVAs totaling 31 hours, 58 minutes were performed. Duration was 114 days, 15 hours, 22 minutes … (USSR).

06/05/1991 … STS-40 Columbia … Bryan O'Connor, Sidney Gutierrez, James Bagian, Tamara Jernigan, Rhea Seddon, Drew Gaffney, Millie Hughes-Fulford … The first Spacelab for Life Sciences (SLS-1); the first dedicated life science research. Duration was 9 days, 02 hours, 15 minutes … (USA).

08/02/1991 … STS-43 Atlantis … John Blaha, Michael Baker, Shannon Lucid, David Low, James Adamson … The Tracking Data Relay Satellite

(TDRS-E) was deployed; Lucid was the first woman to make three space flights. Duration was 8 days, 21 hours, 22 minutes … (USA).

09/12/1991 … STS-48 Discovery … John Creighton, Kenneth Reightler, Charles Gemar, James Buchli, Mark Brown … The Upper Atmosphere Research Satellite (UARS) was deployed. Duration was 5 days, 08 hours, 28 minutes … (USA).

10/02/1991 … Soyuz TM-13 … Aleksandr Volkov, Franz Viehboeck, Toktar Aubakirov … The tenth Mir long-stay mission; two researchers; Viehboeck of Austria and Aubakirov of Kazakhstan returned to Earth on board Soyuz TM-12; Volkov returned on board Soyuz TM-13; one EVA was performed. Duration was 175 days, 02 hours, 52 minutes … (USSR).

11/24/1991 … STS-44 Atlantis … Frederick Gregory, Terrence Henricks, James S. Voss, Story Musgrave, Mario Runco, Thomas Hennen … A classified Department of Defense (DOD) satellite was deployed; contamination research was performed. Duration was 6 days, 22 hours, 52 minutes … (USA).

01/22/1992 … STS-42 Discovery … Ronald Grabe, Stephen Oswald, Norman Thagard, William Readdy, David Hilmers, Roberta Bondar, Ulf Merbold … Spacelab International Microgravity Laboratory (IML-1). Duration was 8 days, 01 hour, 16 minutes … (USA).

03/17/1992 … Soyuz TM-14 … Aleksandr Viktorenko, Aleksandr Kaleri, Klaus-Dietrich Flade … The eleventh Mir long-stay mission; Flade of Germany returned to Earth on board Soyuz TM-13; Viktorenko and Kaleri returned on board Soyuz TM-14; a 123-minute EVA was performed. Duration was 145 days, 14 hours, 11 minutes … As of December, 1991, with the collapse of the Soviet Union, space flights originating out of Russia will be designated as "Russia," instead of "USSR" … (Russia).

03/24/1992 … STS-45 Atlantis … Charles Bolden, Brian Duffy, Kathryn Sullivan, David Leestma, Michael Foale, Dirk Frimout, Byron Lichtenberg … Application Laboratory for Applications and Science (ATLAS-1). Duration was 8 days, 22 hours, 10 minutes … (USA).

05/07/1992 ... STS-49 Endeavor ... Daniel Brandenstein, Kevin Chilton, Bruce Melnick, Pierre Thuot, Richard Hieb, Kathryn Thornton, Thomas Akers ... The first flight of the shuttle Endeavor; four EVAs by four crew members, totaling a duration record of 60.1 hours; the first three-person EVA; Intelsat-VI was recovered from orbit and then redeployed by the crew. Duration was 8 days, 21 hours, 19 minutes ... (USA).

06/25/1992 ... STS-50 Columbia ... Richard Richards, Kenneth Bowersox, Bonnie Dunbar, Ellen Baker, Carl Meade, Lawrence DeLucas, Eugene Trinh ... The U.S. Microgravity Laboratory (USML-1); the first Extended Duration Orbiter (EDO); shuttle duration record. Duration was 13 days, 19 hours, 31 minutes ... (USA).

07/27/1992 ... Soyuz TM-15 ... Anatoliy Solovyov, Sergei Avdeyev, Michael Tognini ... The twelfth Mir long-stay mission; Togini of France returned to Earth on board Soyuz TM-14; Solovyov and Avdeyev returned on board Soyuz TM-15; four EVAs were performed, totaling 18 hours, 21 minutes. Duration was 188 days, 21 hours, 41 minutes ... (Russia).

07/31/1992 ... STS-46 Atlantis ... Loren Shriver, Andrew Allen, Jeffery Hoffman, Marsha Ivins, Franklin Chang-Diaz, Claude Nicollier, Franco Malerba ... The EURECA platform was deployed; tether experiment TSS-1. Duration was 7 days, 23 hours, 16 minutes ... (USA).

09/12/1992 ... STS-47 Endeavor ... Robert Gibson, Curtis Brown, Mark Lee, Jay Apt, Jan Davis, Mamoru Mohri, Mae Jemison ... The 50th shuttle mission; Jemison is the first black woman in space; Mohri is the first Japanese national to fly in space aboard the shuttles; Lee and Davis are the first married couple to fly into space together; first Japanese Spacelab. Duration was 7days, 22 hours, 31 minutes ... (USA).

10/22/1992 ... STS-52 Columbia ... James Wetherbee, Michael Baker, Tamara Jernigan, William Shepherd, Charles Veach, Steven MacLean ... Deployment of the LAGEOS 2; U.S. Microgravity payload (USMP); materials experiments. Duration was 9 days, 20 hours, 57 minutes ... (USA).

12/02/1992 ... STS-53 Discovery ... David Walker, Robert Cabana, Guion Bluford, Michael Clifford, James S. Voss ... The last Department of Defense (DOD) classified mission for the shuttles; a DOD satellite was deployed; laser experimentation. Duration was 7 days, 07 hours, 21 minutes ... (USA).

01/13/1993 ... STS-54 Endeavor ... John Casper, Donald McMonagle, Gregory Harbaugh, Susan Helms, Mario Runco ... Deployment of the TDRS-F; Differential X-ray Experiment (DXS); 268 minute EVA was performed. Duration was 5 days, 23 hours, 39 minutes ... (USA).

01/24/1993 ... Soyuz TM-16 ... Gennadi Manakov, Aleksandr Polishchuk ... The thirteenth Mir long-stay mission; the first docking with the Kristall androgynous port module; two EVAs were performed. Duration was 179 days, 44 minutes ... (Russia).

04/08/1993 ... STS-56 Discovery ... Kenneth Cameron, Stephen Oswald, Kenneth Cockrell, Michael Foale, Ellen Ochoa ... Ochoa is the first Hispanic woman in space; the second Atmospheric Mission (ATLAS-2); SPARTAN-2. Duration was 9 days, 06 hours, 09 minutes ... (USA).

04/26/1993 ... STS-55 Columbia ... Steven Nagel, Terrence Henricks, Charles Precourt, Jerry Ross, Bernard Harris, Hans Schlegel, Ulrich Walter ... The second German Spacelab mission (D2); Earth observation and astronomy experiments; the shuttle program now exceeded one year (aggregate) flight time. Duration was 9 days, 23 hours, 41 minutes ... (USA).

06/21/1993 ... STS-57 Endeavor ... Ronald Grabe, Brian Duffy, David Low, Janice E. Voss, Nancy Sherlock (Currie), Peter Wisoff ... The first Spacehab; EURECA retrieval; GAS; EVA performed. Duration was 9 days, 23 hours, 46 minutes ... (USA).

07/01/1993 ... Soyuz TM-17 ... Vasili Tsiblyev, Aleksandr Serebrov, Jean-Pierre Haignere ... The fourteenth Mir long-stay mission; Haignere of France returned to Earth on board Soyuz TM-16; Tsibliyev and Serebrov returned on board Soyuz TM-17; five EVAs were performed. Duration was 196 days, 17 hours, 45 minutes ... (Russia).

09/12/1993 ... STS-51 Discovery ... Frank Culbertson, William Readdy, Daniel Bursch, James Newman, Carl Walz ... Advanced Comsat ACTS/SPAS was deployed; EVA was completed; the first nighttime Kennedy Space Center landing. Duration was 9 days, 20 hours, 12 minutes ... (USA).

10/18/1993 ... STS-58 Columbia ... John Blaha, Richard Searfoss, Rhea Seddon, William MacArthur, David Wolf, Shannon Lucid, Martin Fettman ... The first dissection was performed in space; the second Spacelab for Life Sciences (SLS-2); the second EDO; new shuttle record duration. Duration was 14 days, 14 minutes ... (USA).

12/02/1993 ... STS-61 Endeavor ... Richard Covey, Kenneth Bowersox, Story Musgrave, Claude Nicollier, Kathryn Thornton, Thomas Akers, Jeffery Hoffman ... The first Hubble Space Telescope servicing mission; five EVAs were performed for a total time of 35 hours, 28 minutes, with four crew members; Akers sets a new U.S. EVA duration record of 29 hours, 40 minutes. Total flight duration was 10 days, 19 hours, 59 minutes ... (USA).

01/08/1994 ... Soyuz TM-18 ... Viktor Afanasyev, Yuri Usachyov, Valeri Polyakov ... The fifteenth Mir long-stay mission; physician Polyakov remained on board Mir for a record duration, returning to Earth on board Soyuz TM-20. Duration was 182 days, 27 minutes ... (Russia).

02/03/1994 ... STS-60 Discovery ... Charles Bolden, Kenneth Reightler, Sergei Krikalyov, Jan Davis, Franklin Chang-Diaz, Ronald Sega ... Krikalyov was the first Russian cosmonaut on board an American shuttle; the attempt to deploy the Wake Shield Facility (a device to create vacuums in space) failed; Spacehab 2. Duration was 8 days, 07 hours, 10 minutes ... (USA).

03/04/1994 ... STS-62 Columbia ... John Casper, Andrew Allen, Pierre Thuot, Charles Gemar, Marsha Ivins ... Microgravity Payload (USMP-2); OAST-2; SSBUV/A; DEE; the third EDO. Duration was 13 days, 23 hours, 18 minutes ... (USA).

04/09/1994 ... STS-59 Endeavor ... Sidney Gutierrez, Kevin Chilton, Linda Godwin, Jay Apt, Michael Clifford, Thomas Jones ... The first flight of the Space Radar Lab; mapped 20% of the Earth's surface in three dimensions; environmental studies were conducted; CONCAP-IV; GAS; NIH-Experiments; the shuttle Endeavor made a record 412 maneuvers. Duration was 11 days, 05 hours, 50 minutes ... (USA).

07/01/1994 ... Soyuz TM-19 ... Yuri Malenchenko, Talgat Musabayev ... The sixteenth Mir long-stay mission; two EVAs were conducted. Duration was 125 days, 22 hours, 54 minutes ... (Russia).

07/08/1994 ... STS-65 Columbia ... Robert Cabana, James Halsell, Richard Hieb, Leroy Chiao, Donald Thomas, Chiaki Mukai, Carl Walz ... Two-day flight of the International Microgravity Laboratory (IML-2); more than eighty experiments were conducted; the fourth EDO, with a record shuttle duration. Duration was 14 days, 17 hours, 56 minutes ... (USA).

09/09/1994 ... STS-64 Discovery ... Richard Richards, Blaine Hammond, Jerry Linenger, Carl Meade, Mark Lee, Susan Helms ... Laser experimentation LITE; SPARTAN-3; SAFER (EVA); ROMPS. Duration was 10 days, 22 hours, 51 minutes ... (USA).

09/30/1994 ... STS-68 Endeavor ... Michael Baker, Terrence Wilcutt, Thomas Jones, Steven Smith, Peter Wisoff, Daniel Bursch ... Space Radar Lab SRL-2; GAS; Stamps; secondary payloads. Duration was 11 days, 05 hours, 47 minutes ... (USA).

10/04/1994 ... Soyuz TM-20 ... Aleksandr Viktorenko, Yelena Kondakova, Ulf Merbold ... The seventeenth Mir long-stay mission; Merbold (ESA) returned to Earth on board Soyuz TM-19. Duration was 169 days, 05 hours, 22 minutes ... (Russia).

11/03/1994 ... STS-66 Atlantis ... Donald McMonagle, Curtis Brown, Scott Parazynski, Ellen Ochoa, Jean-Francis Clervoy, Joseph Tanner ... Atmospheric Mission ATLAS-3; CRISTA/SPAS; ESCAPE-2. Duration was 10 days, 22 hours, 35 minutes ... (USA).

02/03/1995 ... STS-63 Discovery ... James Wetherbee, Eileen Collins, Vladimir Titov, Michael Foale, Janice E. Voss, Bernard Harris ... Rendezvous with the Russian Mir; the first female shuttle pilot (Collins); the second Russian cosmonaut (Titov); SPARTAN-4; EVA. Duration was 8 days, 06 hours, 30 minutes ... (USA).

03/02/1995 ... STS-67 Endeavor ... Stephen Oswald, William Gregory, Wendy Lawrence, Ronald Parise, Samuel Durrance, Tamara Jernigan, John Grunsfeld ... The second UV Astronomy Spacelab; the fifth EDO; a record duration for the shuttle program to date. Duration was 16 days, 15 hours, 10 minutes ... (USA).

03/14/1995 ... Soyuz TM-21 ... Vladimir Dezhurov, Gennadi Strekalov, Norman Thagard ... The eighteenth Mir long-stay mission; Thagard was the first NASA researcher to stay on board Mir; all the crew members returned to Earth on board shuttle flight STS-71; three EVAs were performed. Duration was 181 days, 41 minutes ... (Russia).

06/27/1995 ... STS-71 Atlantis ... Robert Gibson, Charles Precourt, Ellen Baker, Gregory Harbaugh, Bonnie Dunbar, Anatoliy Solovyov, Nikolai Budarin ... The first Mir docking with an exchange of flight crews; Spacelab was carried on board; Vladimir Dezhurov, Gennadi Strekalov and Norman Thagard returned to Earth on board STS-71. Duration was 9 days, 19 hours, 23 minutes ... (USA).

07/13/1995 ... STS-70 Discovery ... Terrence Henricks, Kevin Kregel, Donald Thomas, Nancy Currie (Sherlock), Mary Weber ... Deployed the TDRS-G; crystal growth and biological experiments. Duration was 8 days, 22 hours, 21 minutes ... (USA).

09/03/1995 ... Soyuz TM-22 ... Yuri Gidzenko, Sergei Avdeyev, Thomas Reiter ... The nineteenth Mir long-stay mission; ESA researcher Reiter performed two EVAs. Duration was 179 days, 01 hour, 42 minutes ... (Russia).

09/07/1995 ... STS-69 Endeavor ... David Walker, Kenneth Cockrell, James S. Voss, James Newman, Michael Gernhardt ... The second flight/experiment of the Wake Shield Facility; the 30th shuttle EVA. Duration was 10 days, 20 hours, 30 minutes … (USA).

10/20/1995 … STS-73 Columbia … Kenneth Bowersox, Kent Rominger, Kathryn Thornton, Catherine Coleman, Albert Sacco, Michael Lopez-Alegria, Fred Leslie ... Spacelab USML-2; educational experiments; the sixth shuttle EDO. Duration was 15 days, 21 hours, 53 minutes … (USA).

11/12/1995 … STS-74 Atlantis … Kenneth Cameron, James Halsell, Jerry Ross, William McArthur, Chris Hadfield ... Docked with the Mir orbiting platform; delivered a docking unit for future shuttle missions; delivered a new solar array. Duration was 8 days, 04 hours, 32 minutes … (USA).

01/11/1996 … STS-72 Endeavor … Brian Duffy, Brent Jett, Leroy Chiao, Daniel Barry, Winston Scott, Koichi Wakata ... Retrieved the orbiting SFU Space Flyer Unit; SPARTAN/OAST flyer; two EVAs were performed. Duration was 8 days, 22 hours, 02 minutes … (USA).

02/21/1996 … Soyuz TM-23 … Yuri Onufrienko, Yuri Usachyov ... A Mir long-stay mission; six EVAs were performed, totaling 30 hours, 31 minutes. Duration was 193 days, 19 hours, 08 minutes … (Russia).

02/22/1996 … STS-75 Columbia … Andrew Allen, Scott Horowitz, Jeffery Hoffman, Franklin Chang-Diaz, Umberto Guidoni, Maurizio Cheli, Claude Nicollier ... Microgravity Payload USMP-3; Tether Satellite TSS-1R; OARE; the seventh shuttle EDO mission. Duration was 15 days, 17 hours, 41 minutes … (USA).

03/22/1996 … STS-76 Atlantis … Kevin Chilton, Richard Searfoss, Shannon Lucid, Linda Godwin, Ronald Sega, Michael Clifford ... Docked with the Mir; Spacehab short module was delivered; delivered Shannon Lucid for an extended stay on board the Mir; EVA mounted experiments on Mir's docking module. Duration was 9 days, 05 hours, 17 minutes … (USA).

05/19/1996 ... STS-77 Endeavor ... John Casper, Curtis Brown, Daniel Bursch, Mario Runco, Marc Garneau, Andrew Thomas ... Inflatable Antenna Experiment (IAE); Spacehab 4; PAMS; SPARTAN; TEAMS. Duration was 10 days, 40 minutes ... (USA).

06/20/1996 ... STS-78 Columbia ... Terrence Henricks, Kevin Kregel, Susan Helms, Richard Linnehan, Charles Brady, Jean-Jacques Favier, Robert Thirsk ... Life and Microgravity Science Spacelab (LMS); SAREX-II; the eighth EDO shuttle mission; shuttle duration record set to date. Duration was 16 days, 21 hours, 49 minutes ... (USA).

08/17/1996 ... Soyuz TM-24 ... Valeri Korzun, Aleksandr Kaleri, Claudie Andre-Deshays ... Mir long-stay mission; Andre-Deshays of France returned to Earth on board Soyuz TM-23; two EVAs were performed. Duration was 196 days, 17 hours, 26 minutes ... (Russia).

09/16/1996 ... STS-79 Atlantis ... William Readdy, Terrence Wilcutt, Thomas Akers, Jay Apt, John Blaha, Carl Walz ... Docked with Mir; Spacehab double module; crew exchange (Blaha and Lucid); Lucid set a new U.S. space endurance record of 188 days, 4 hours; an EVA was performed. Flight duration was 10 days, 03 hours, 20 minutes ... (USA).

11/19/1996 ... STS-80 Columbia ... Kenneth Cockrell, Kent Rominger, Tamara Jernigan, Thomas Jones, Story Musgrave ... The third test of the Wake Shield Facility; ORFEUS-SPAS II; Space Experiment Module (SEM); an EVA was performed; the ninth shuttle EDO mission. Duration was 17 days, 15 hours, 54 minutes ... (USA).

01/12/1997 ... STS-81 Atlantis ... Michael Baker, Brent Jett, Peter Wisoff, John Grunsfeld, Marsha Ivins, Jerry Linenger ... Docked with the orbiting Mir; Spacehab-DM; SAREX-II; crew exchange (Linenger and Blaha ... with John Blaha returning to Earth on board STS-81). Duration was 10 days, 04 hours, 56 minutes ... (USA).

02/10/1997 ... Soyuz TM-25 ... Vasili Tsibliyev, Aleksandr Lazutkin, Reinhold Ewald ... A Mir long-stay mission; Ewald of Germany returned to Earth on board Soyuz TM-24; there was a near-catastrophic collision with the unmanned Progress resupply vessel on 06/25/1997; one EVA was conducted. Duration was 184 days, 22 hours, 08 minutes … (Russia).

02/11/1997 … STS-82 Discovery … Kenneth Bowersox, Scott Horowitz, Joseph Tanner, Steven Hawley, Gregory Harbaugh, Marl Lee, Steven Smith … The second Hubble Space Telescope servicing mission; five EVAs were performed, with four crew members, totaling 33 hours, 11 minutes; ten instrumentation packages were replaced on the telescope. Duration was 9 days, 23 hours, 38 minutes … (USA).

04/04/1997 … STS-83 Columbia … James Halsell, Susan Still, Janice Voss, Michael Gernhardt, Donald Thomas, Roger Crouch, Greg Linteris … Microgravity Science Laboratory (MSL-1), but returned 12 days early due to a fuel cell problem; the re-flight mission was then scheduled for STS-94. Duration was 3 days, 23 hours, 14 minutes … (USA).

05/15/1997 … STS-84 Atlantis … Charles Precourt, Eileen Collins, Jean-Francois Clervoy, Carlos Noriega, Edward Lu, Yelena Kondakova, Michael Foale … The sixth Mir docking for the shuttles; Spacehab-DM; crew exchange (Linenger for Foale). Duration was 9 days, 05 hours, 21 minutes … (USA).

07/01/1997 … STS-94 Columbia … James Halsell, Susan Still, Janice Voss, Michael Gernhardt, Donald Thomas, Roger Crouch, Greg Linteris … Microgravity Science Laboratory (MSL-1) re-flight mission with the same crew from STS-83. Duration was 15 days, 16 hours, 46 minutes … (USA).

08/05/1997 … Soyuz TM-26 … Anatoliy Solovyov, Pavel Vinogradov … Mir long-stay mission; seven EVAs performed to repair the damage to Mir's solar arrays caused by the prior collision with a Progress unmanned resupply vessel; French astronaut Leopold Eyharts returned to Earth on board this Soyuz vessel. Duration was 197 days, 17 hours, 35 minutes … (Russia).

08/07/1997 … STS-85 Discovery … Curtis Brown, Kent Rominger, Jan Davis, Robert Curbeam, Stephen Robinson, Bjarni Tryggvason … CRISTA/SPAS II; Japanese Manipulator (MFD); small payloads. Duration was 11 days, 20 hours, 28 minutes … (USA).

09/25/1997 ... STS-86 Atlantis ... James Wetherbee, Michael Bloomfield, Scott Parazynski, Vladimir Titov, Jean-Loup Chretien, Wendy Lawrence, David Wolf ... The seventh Mir docking for the shuttles; Spacehab-DM; crew exchange (Foale for Wolf, with Wolf returning to Earth on board this shuttle flight). Duration was 10 days, 19 hours, 22 minutes ... (USA).

11/19/1997 ... STS-87 Columbia ... Kevin Kregel, Steve Lindsey, Kalpana Chawla, Winston Scott, Takao Doi, Leonid Kadenyuk ... Microgravity Payload (USMP-4); SPARTAN-201; two EVAs were performed; small payloads. Duration was 15 days, 16 hours, 35 minutes ... (USA).

01/23/1998 ... STS-89 Endeavor ... Terrence Wilcutt, Joe Edwards, James Reilly, Michael Anderson, Bonnie Dunbar, Salizhan Sharipov, Andrew Thomas ... The eighth Mir docking for the shuttles; Spacehab-DM; crew exchange (Wolf for Thomas, with David Wolf returning to Earth on board this shuttle flight). Duration was 8 days, 19 hours, 48 minutes ... (USA).

01/29/1998 ... Soyuz TM-27 ... Talgat Musabayev, Nikolai Budarin, Leopold Eyharts ... Mir long-stay mission; Eyharts of France would return to Earth on board Soyuz TM-26; Yuri Baturin returned on board this Soyuz flight; five EVAs were performed. Duration was 207 days, 12 hours, 51 minutes ... (Russia).

04/17/1998 ... STS-90 Columbia ... Richard Searfoss, Scott Altman, Kathryn Hire, Richard Linnehan, Dafydd Williams, Jay Buckey, James Pawelczyk ... Neurolab (the 16th Spacelab); bioreactor; small payloads (GAS). Duration was 15 days, 21 hours, 51 minutes ... (USA).

06/02/1998 ... STS-91 Discovery ... Charles Precourt, Dominic Gorie, Wendy Lawrence, Franklin Chang-Diaz, Janet Kavandi, Valeri Ryumin ... Docked with the Mir; Spacehab-SM; Andrew Thomas returned to Earth from Mir on board this shuttle flight; first GAS payload. Duration was 9 days, 19 hours, 55 minutes ... (USA).

08/13/1998 ... Soyuz TM-28 ... Gennadi Padalka, Sergei Avdeyev, Yuri Baturin ... Mir long-stay mission; Baturin would return to Earth on board Soyuz TM-27; Ivan Bella would return on board this Soyuz vessel; Avdeyev remained on board Mir; one EVA was performed. Duration was 198 days, 16 hours, 31 minutes ... (Russia).

10/29/1998 ... STS-95 Discovery ... Curtis Brown, Steve Lindsey, Scott Parazynski, Stephen Robinson, Pedro Duque, Chiaki Mukai, John Glenn ... Spacehab-SM; SPARTAN-201; John Glenn, the first American to orbit the Earth on board Mercury-Atlas 6 on February the 20th, 1962, returns to space on this mission 36 years later. Duration was 8 days, 21 hours, 45 minutes ... (USA).

12/04/1998 ... STS-88 Endeavor ... Robert Cabana, Frederick Sturckow, Nancy Currie, Jerry Ross, James Newman, Sergei Krikalyov ... The first International Space Station (ISS) assembly flight 2A; Unity Module; three EVAs with two crew members totaling 21 hours, 22 minutes; the first habitation of the ISS. Duration was 11 days, 19 hours, 19 minutes ... (USA).

02/20/1999 ... Soyuz TM-29 ... Victor Afanasyev, Jean-Pierre Haignere, Ivan Bella ... Mir long-stay mission; Bella returned to Earth on board Soyuz TM-28; Sergei Avdeyev returned on board this Soyuz vessel; three EVAs were performed. Duration was 188 days, 20 hours, 16 minutes ... (Russia).

05/27/1999 ... STS-96 Discovery ... Kent Rominger, Rick Husband, Ellen Ochoa, Tamara Jernigan, Daniel Barry, Julie Payette, Valeri Tokarev ... The second ISS assembly flight 2A.1, Spacehab-DM, Starshine; the first docking with the ISS; one EVA was performed with two crew members totaling seven hours, 55 minutes. Duration was 9 days, 19 hours, 14 minutes ... (USA).

07/23/1999 ... STS-93 Columbia ... Eileen Collins, Jeffery Ashby, Steven Hawley, Catherine Coleman, Michel Tognini ... Deployed the Chandra/AXAF; first female shuttle commander (Collins). Duration was 4 days, 22 hours, 50 minutes ... (USA).

12/20/1999 ... STS-103 Discovery ... Curtis Brown, Scott Kelly, Steven Smith, Michael Foale, John Grunsfeld, Claude Nicollier, Jean-Francois Clervoy ... The third Hubble Space Telescope servicing mission; three EVAs were performed with four crew members, totaling 24 hours, 33 minutes. Duration was 7 days, 23 hours, 12 minutes ... (USA).

02/11/2000 ... STS-99 Endeavor ... Kevin Kregel, Dominic Gorie, Janet Kavandi, Janice Voss, Mamoru Mohri, Gerhard Thiele ... Shuttle Radar Topography Mission (SRTM); used radar systems to obtain a high-resolution digital topographic database of the Earth's surface. Duration was 11 days, 05 hours, 40 minutes ... (USA).

04/04/2000 ... Soyuz TM-30 ... Sergei Zalyetin, Aleksandr Kalen ... The final Mir long-stay mission; reactivated Mir under contract with MirCorp (a private concern). Duration was 72 days, 19 hours, 42 minutes ... (Russia).

05/19/2000 ... STS-101 Atlantis ... James Halsell, Scott Horowitz, Mary Weber, Jeffrey Williams, James Voss, Susan Helms, Yuri Usachyov ... The third ISS shuttle assembly flight 2A.2a; Spacehab-DM; one EVA was performed using two crew members, totaling 6 hours, 44 minutes. Duration was 9 days, 20 hours, 10 minutes ... (USA).

09/08/2000 ... STS-106 Atlantis ... Terrence Wilcut, Scott Altman, Daniel Burbank, Edward Lu, Richard Mastracchio, Yuri Malenchenko, Boris Morukov ... The fourth ISS assemble flight 2A.2b; Spacehab-DM; One EVA was performed with two crew members, totaling 6 hours, 14 minutes. Duration was 11 days, 19 hours, 12 minutes ... (USA).

10/11/2000 ... STS-92 Discovery ... Brian Duffy, Pamela Melroy, Koichi Wakata, Leory Chiao, Peter Wisoff, Michael Lopez-Alegria, William McArthur ... The fifth ISS assembly flight 3A; PMA-3; IMAX; four EVAs were conducted with four crew members, totaling 27 hours, 19 minutes. Duration was 12 days, 21 hours, 44 minutes ... (USA).

10/31/2000 ... Soyuz TM-31 ... Sergei Krikalyov, Yuri Gidzenko, William Shepherd ... Delivered the ISS Expedition 1 crew (Shepherd, Gidzenko and Krikalyov); first ISS extended stay totaling 136 days on the

ISS; the three crew members returned to Earth on board STS-102; Talgat Musabayev, Yuri Baturin and Dennis Tito returned to Earth on board this Soyuz vessel. Duration was 186 days, 21 hours, 49 minutes … (Russia).

12/01/2000 … STS-97 Endeavor … Brent Jett, Michael Bloomfield, Joseph Tanner, Carlos Noriega, Marc Garneau ... The sixth ISS assembly flight 4A; PV Module P6; three EVAs were carried out with two flight crew members, totaling 19 hours, 20 minutes. Duration was 10 days, 19 hours, 58 minutes … (USA).

02/07/2001 … STS-98 Atlantis … Kenneth Cockrell, Mark Polansky, Robert Curbeam, Thomas Jones, Marsha Ivins ... The seventh ISS assembly flight 5A; U.S. Lab; three EVAs were conducted with two crew members, totaling 19 hours, 49 minutes. Duration was 12 days, 21 hours, 21 minutes … (USA).

03/08/2001 … STS-102 Discovery … James Wetherbee, James Kelly, Andrew Thomas, Paul Richards, Yuri Usachyov, James Voss, Susan Helms ... The eight ISS assembly flight 5A; Leonardo MPLM; delivered the Expedition 2 crew of Helms, Voss and Usachyov; returned to Earth the Expedition 1 crew of William Shepherd, Yuri Gidzenk and Sergei Krikalyov; two EVAs were conducted with four crew members, totaling 15 hours, 26 minutes. Duration was 12 days, 19 hours, 51 minutes … (USA).

04/19/2001 … STS-100 Endeavor … Kent Rominger, Jeffrey Ashby, Chris Hadfield, Scott Parazynski, John Phillips, Umberto Guidono, Yuri Lonchakov ... The ninth ISS assembly flight 6A; Canadarm-2, robotic arm was delivered to the ISS; Raffaello MPLM; two EVAs were conducted by two crew members, totaling 14 hours, 50 minutes. Duration was 11 days, 21 hours, 31 minutes … (USA).

04/28/2001 … Soyuz TM-32 … Talgat Musabayev, Yuri Baturin, Dennis Tito ... Exchanged the Soyuz TM-32 for the TM-31 to serve the ISS as an emergency escape vessel; Tito was the first space "tourist," having paid approximately twenty million dollars to the Russians for the privilege of riding into orbit; Victor Afanasyev, Konstantin Kozeyev and Claudie Haignere returned to Earth with the vessel. Duration was 185 days, 21 hours, 23 minutes … (Russia).

07/12/2001 ... STS-104 Atlantis ... Steve Lindsey, Charles Hobaugh, Michael Gernhardt, James Reilly, Janet Kavandi ... The tenth ISS assembly flight 7A; ISS Airlock delivered; three EVAs were conducted by two crew members, totaling 16 hours, 30 minutes. Duration was 12 days, 18 hours, 37 minutes ... (USA).

08/10/2001 ... STS-105 Discovery ... Scott Horowitz, Frederick Sturckow, Daniel Barry, Patrick Forrester, Frank Culbertson, Vladimir Dezhurov, Mikhail Tyurin ... The eleventh ISS assembly flight; Leonardo MPLM; delivered the Expedition 3 crew members (Culbertson, Dezhurov and Tyurin) to the ISS; returned the Expedition 2 crew to Earth (James Voss, Susan Helms and Yuri Usachyov); two EVAs were conducted by two crew members, totaling 11 hours, 45 minutes. Duration was 11 days, 21 hours, 14 minutes ... (USA).

10/21/2001 ... Soyuz TM-33 ... Victor Afanasyev, Konstantin Kozeyev, Claudie Haignere ... Exchanged Soyuz TM-33 for Soyuz TM-32 to serve the ISS as an emergency escape vehicle; Yuri Gidzenko, Roberto Vittori and Marc Shuttleworth would return to Earth on board TM-33. Duration was 195 days, 18 hours, 52 minutes ... (Russia).

12/05/2001 ... STS-108 Endeavor ... Dominic Gorie, Mark Kelly, Linda Godwin, Daniel Tani, Yuri Onufrienko, Carl Walz, Daniel Bursch ... ISS utilization flight UF-1; Raffaello MPLM, GAS, MACH-1; delivered Expedition 4 crew (Bursch, Walz and Onufrienko); returned to Earth the Expedition 3 crew of Frank Culbertson, Vladimir Dezhurov and Michael Tyurin; one EVA was conducted with two crew members, totaling 4 hours, 11 minutes. Duration was 11 days, 19 hours, 37 minutes ... (USA).

03/01/2002 ... STS-109 Columbia ... Scott Altman, Duane Carey, John Grunsfeld, Nancy Currie, James Newman, Richard Linnehan, Michael Massimino ... The fourth Hubble Space Telescope servicing mission; Advanced Camera for Surveys (ACS); five EVAs were conducted with four crew members, totaling 35 hours, 55 minutes. Duration was 10 days, 22 hours, 11 minutes ... (USA).

04/08/2002 ... STS-110 Atlantis ... Michael Bloomfield, Stephen Frick, Jerry Ross, Steven Smith, Ellen Ochoa, Lee Morin, Rex Walheim ... ISS assemble flight 8A; Center Integrated Truss Assembly S0 (ITS S0); Mobile Transporter (MT); four EVAs were conducted with four crew members, totaling 28 hours, 22 minutes. Duration was 10 days, 19 hours, 44 minutes ... (USA).

04/25/2002 ... Soyuz TM-34 ... Yuri Gidzenko, Roberto Vittori, Mark Shuttleworth ... Exchanged the Soyuz TM-34 for TM-33 to serve as an emergency escape vessel for the ISS; Shuttleworth was the first South African to go into space, and the second space tourist, after paying the Russians in excess of twenty million dollars for the privilege; Sergei Zalyotin, Frank DeWinne and Yuri Lonchakov returned to Earth on this vessel. Duration was 198 days, 17 hours, 38 minutes ... (Russia).

06/05/2002 ... STS-111 Endeavor ... Kenneth Cockrell, Paul Lockhart, Franklin Chang-Diaz, Philippe Perrin, Valeri Korzun, Sergei Treschev, Peggy Whitson ... ISS utilization flight UF-2; Leonardo MPLM; Mobile Base Systems (MBS); delivered the Expedition 5 crew (Whitson, Treschev and Korzun), and returned to Earth the Expedition 4 crew of Daniel Bursch, Carl Walz and Yuri Onufrienko; three EVAs were conducted with two crew members, totaling 19 hours, 31 minutes. Duration was 13 days, 20 hours, 36 minutes ... (USA).

10/07/2002 ... STS-112 Atlantis ... Jeffery Ashby, Pamela Melroy, David Wolf, Piers Sellers, Sandra Magnus, Fyodor Yurchikhin ... ISS assembly flight 9A; Integrated Truss Assembly S1 (ITS S1); Crew Equipment Translation Aid (CETA); three EVAs were conducted with two crew members, totaling 19 hours, 41 minutes. Duration was 10 days, 19 hours, 59 minutes ... (USA).

10/30/2002 ... Soyuz TMA-1 ... Sergei Zalyotin, Frank De Winne, Yuri Lonchakov ... Exchanged the Soyuz TMA-1 for TM-34 to serve the ISS as an emergency escape vessel; returned to Earth the Expedition 6 crew of Donald Pettit, Kenneth Bowersox and Nikolai Budarin. Duration was 185 days, 22 hours, 56 minutes ... (Russia).

11/24/2002 ... STS-113 Endeavor ... James Wetherbee, Paul Lockhart, Michael Lopez-Alegria, John Herrington, Kenneth Bowersox, Nikolai Budarin, Donald Pettit ... ISS assembly flight 11A; Integrated Truss

Assembly P1 (ITA P1); CETA; delivered the Expedition 6 crew (Bowersox, Budarin and Pettit); returned to Earth the Expedition 5 crew of Peggy Whitson, Sergei Treschev and Valeri Korzun; three EVAs were conducted by two crew members, totaling 19 hours, 55 minutes. Duration was 13 days, 18 hours, 49 minutes … (USA).

02/01/2003 … STS-107 Columbia … Rick D. Husband, William C. McCool, Michael P. Anderson, Kalpana Chawla, David Brown, Laurel Clark, Ilan Ramon ... Spacehab-DM; Freestar; Ramon was the first Israeli astronaut in space; the orbiter broke up during the reentry phase, resulting in the loss of the crew and the vehicle; the investigation would reveal that leading-edge heat shield tiles on the port wing were damaged during liftoff, which in turn resulted in a catastrophic fire-related event, leading to the destruction of the shuttle during the reentry into the Earth's atmosphere. Duration was 15 days, 22 hours, 20 minutes … (USA).

04/26/2003 … Soyuz TMA-2 … Yuri Malenchenko, Edward Lu, Pedro Duque .,. Exchanged the Soyuz TMA-2 for the Soyuz TMA-1 to serve the ISS as an emergency escape vessel; delivered and returned to Earth the Expedition 7 crew of Lu and Malenchenko. Duration was 183 days, 22 hours, 47 minutes … (Russia).

10/15/2003 … Shenzhou 5 … Yang Liwei … The first Chinese human space flight after the two failed attempts in 1974 and 1975; completed 14 orbits of the Earth. Duration was 21 hours, 23 minutes … (China).

10/18/2003 … Soyuz TMA-3 … Aleksandr Kaleri, Michael Foale Pedro Duque ... Exchanged the Soyuz TMA-3 for the TMA-2 to serve as an emergency escape vessel for the ISS; delivered and returned the Expedition 8 crew of Foale and Kaleri. Duration was 194 days, 18 hours, 33 minutes … (Russia).

04/19/2004 … Soyuz TMA-4 … Gennadi Padalka, E. Michael Fincke, Andre Kuipers ... Exchanged the Soyuz TMA-4 for the TMA-3 to serve the ISS as an emergency escape vessel; delivered and returned the Expedition 9 crew of Padalka and Fincke. Duration was 187 days, 21 hours, 16 minutes … (Russia).

06/21/2004 ... SpaceShipOne Flight 15 ... Michael Melvill ... The first privately funded non-government manned space flight; a sub-orbital flight; reached an altitude of 328,491 feet (100,124 meters or 62.2 miles). Duration was 24 minutes, 05 seconds ... (USA).

09/29/2004 ... SpaceShipOne Flight 16 ... Michael Melvill ... The second space flight for the vessel, to claim the Ansari X prize; a sub-orbital flight; reached an altitude of 337,569 feet (102,891 meters or 63.9 miles). Duration was 24 minutes, 11 seconds ... (USA).

10/04/2004 ... SpaceShipOne Flight 17 ... Brian Binnie ... Claimed the Ansari X prize; reached a sub-orbital altitude of 367,442 feet (111,997 meters or 69.6 miles). Duration was 23 minutes, 56 seconds ... (USA).

10/14/2004 ... Soyuz TMA-5 ... Saliszan Sharipov, Leroy Chiao, Yuri Shargin ... Exchanged Soyuz TMA-5 for TMA-4 to serve the ISS as an emergency escape vessel; delivered and returned to Earth the Expedition 10 crew of Chiao and Sharipov; Roberto Vittori returned in this vessel. Duration was 192 days, 19 hours, 02 minutes ... (Russia).

04/15/2005 ... Soyuz TMA-6 ... Sergei Krikalyov, John Phillips, Roberto Vittori ... Exchanged Soyuz TMA-6 for TMA-5 to serve as an emergency escape vessel for the ISS; delivered Expedition 11 crew of Krikalyov and Phillips; Gregory Olsen would return to Earth on board this vessel. Duration was 179 days, 23 minutes ... (Russia).

07/26/2005 ... STS-114 Discovery ... Eileen Collins, James Kelly, Charles Camarda, Wendy Lawrence, Soichi Noguchi, Stephen Robinson, Andrew Thomas ... ISS logistics flight (LF-1); Raffaello MPLM; External Stowage Platform; test and evaluation of new safety procedures following the STS-107 catastrophe; three EVAs were performed by two crew members, totaling 20 hours, 05 minutes. Duration was 13 days, 21 hours, 33 minutes ... (USA).

10/01/2005 ... Soyuz TMA-7 ... Valeri Tokarev, William McArthur, Gregory Olsen ... Exchanged Soyuz TMA-7 for TMA-6 to serve as an emergency escape vessel for the ISS; delivered the Expedition 12 crew of McArthur and Tokarev; Olsen was the third paying space tourist; returned

Macros Pontes to Earth in this vessel. Duration was 189 days, 19 hours, 53 minutes … (Russia).

10/12/2005 … Shenzhou 6 … Fei Junlong, Nie Haishen ... The second Chinese human space flight after the two failed attempts in 1974 and 1975; first two man Chinese crew; 76 orbits of the Earth. Duration was 4 days, 19 hours, 32 minutes … (China).

03/30/2006 … Soyuz TMA-8 … Pavel Vinogradov, Jeffery Williams, Macros Pontes ... Exchanged the Soyuz TMA-8 for TMA-7 to serve the ISS as an emergency escape vessel; delivered the Expedition 13 crew of Williams and Vinogradov; Anousheh Ansari, the first female space tourist, would return to Earth on board TMA-8. Duration was 182 days, 22 hours, 43 minutes … (Russia).

07/04/2006 … STS-121 Discovery … Steve Lindsey, Mark Kelly, Mike Fossum, Lisa Nowak, Stephanie Wilson, Piers Sellers, Thomas Reiter ... ISS assembly flight ULF-1.1; Leonardo MPLM; performed ISS maintenance, delivered supplies and crew member Reiter; tested new safety equipment and procedures; three EVAs were conducted with two crew members, totaling 21 hours, 29 minutes. Duration was 12 days, 18 hours, 37 minutes … (USA).

09/09/2006 …STS-115 Atlantis … Brent Jett, Christopher Ferguson, Joseph Tanner, Daniel Burbank, Steven MacLean, Heidemarie Stefanyshyn-Piper ... ISS assembly flight 12A; installed the P3/P4 integrated truss and the second set of solar arrays with rotary joints; three EVAs were conducted with four crew members, totaling 20 hours, 19 minutes. Duration was 11 days, 19 hours, 07 minutes … (USA).

09/18/2006 … Soyuz TMA-9 … Michael Tyurin, Michael Lopez-Alegria, Anousheh Ansari ... Exchanged the Soyuz TMA-9 for TMA-8 to serve the ISS as an emergency escape vessel; delivered the Expedition 14 crew of Lopez-Alegria and Tyurin. Ansari was the first female space tourist, who returned to Earth on board TMA-8. Duration was 215 days, 08 hours, 23 minutes … (Russia).

12/09/2006 ... STS-116 Discovery ... Mark Polansky, William Oefelein, Robert Curbeam, Joan Higginbotham, Nicholas Patrick, Christer Fuglesang, Sunita Williams ... ISS assembly flight 12A.1; installed the P5 integrated truss; rewired the ISS power system; Spacelab-SM; exchanged crew members (Williams for Thomas Reiter); four EVAs were conducted by three crew members, with Curbeam in all four, for a total time of 25 hours, 45 minutes; Fuglesang was the first Swede in space. Duration was 12 days, 20 hours, 45 minutes ... (USA).

04/07/2007 ... Soyuz TMA-10 ... Fyodor Yurchikhin, Oleg Kotov, Charles Simonyi ... Exchanged Soyuz TMA-10 for TMA-9 to serve the ISS as an emergency escape vessel; delivered the Expedition 15 crew (Yurchikhim and Kotov); Simonyi was the fifth space tourist; Sheikh Muszaphar Shukor returned to Earth on board TMA-9. Duration was 196 days, 07 hours, 05 minutes ... (Russia).

06/08/2007 ... STS-117 Atlantis ... Frederick Sturchow, Lee Archambault, Patrick Forrester, Steven Swanson, John Olivas, James Reilly, Clayton Anderson ... ISS assembly flight 13A; installed the S3/S4 integrated truss and the third set of solar arrays with rotary joints; exchanged crew members (Anderson and Williams); four EVAs were conducted by four crew members for a total time of 27 hours, 58 minutes; Sunita Williams returned to Earth on board STS-117. Duration was 13 days, 20 hours, 12 minutes ... (USA).

08/08/2007 ... STS-118 Endeavor ... Scott Kelly, Charles Hobaugh, Tracy Caldwell, Richard Mastracchio, Dafydd Williams, Barbara Morgan, B. Alvin Drew ... ISS assembly flight 13A1; installed the S5 integrated truss; deployed ESP-3; replaced CMG-3; the first use of a station-to-shuttle power transfer system (SSPTS); four EVAs were conducted with two crew members, and one ISS crew member, totaling 23 hours, 15 minutes. Duration was 12 days, 17 hours, 56 minutes ... (USA).

10/10/2007 ... Soyuz TMA-11 ... Yuri Malenchenko, Peggy Whitson, Sheikh Muszaphar Shukor ... Exchanged the Soyuz TMA-11 for TMA-10 to serve as an emergency escape vessel for the ISS; delivered the Expedition 16 crew of Whitson and Malenchenko; Shukor was the first Malaysian in space; Whitson was the first female ISS commander; Yi So-yeon returned to Earth on board TMA-11. Duration was 191 days, 19 hours, 07 minutes … (Russia).

10/23/2007 … STS-120 Discovery … Pamela Melroy, George Zamka, Scott Parazynski, Stephanie Wilson, Douglas Wheelock, Paolo Nespoli, Daniel Tani … ISS assembly flight 10A; installed the Harmony Node 2 module; relocated the P6 integrated truss; exchanged crew members (Tani for Clayton Anderson); four EVAs were conducted by three crew members, totaling 27 hours, 14 minutes. Duration was 15 days, 02 hours, 24 minutes … (USA).

02/07/2008 … STS-122 Atlantis … Stephen Frick, Alan Poindexter, Leland Melvin, Rex Walheim, Hans Schlegel, Stanley Love, Leopold Eyharts … ISS assembly flight 1E; delivered the ESA's Columbus Laboratory; exchanged crew members (Eyharts for Daniel Tani); three EVAs were conducted by three crew members, totaling 22 hours, 08 minutes. Duration was 12 days, 18 hours, 22 minutes … (USA).

03/11/2008 … STS-123 Endeavor … Dominic Gorie, Gregory Johnson, Robert Behnken, Michael Foreman, Richard Linnehan, Takao Doi, Garrett Reisman … ISS assembly flight 1J/A; delivered the Japanese Kibo Experiment Logistics Module and the Canadian Dextre Robotics System; the first full utilization of the station-to-shuttle power transfer system (SSPTS); exchanged crew members (Reisman for Leopold Eyharts); five EVAs were conducted with four crew members, totaling 33 hours, 28 minutes. Duration was 15 days, 18 hours, 12 minutes … (USA).

04/08/2008 … Soyuz TMA-12 … Sergei Volkov, Oleg Kononenko, Yi So-yeon … Exchanged the TMA-12 for the TMA-11 to serve the ISS as an emergency escape vessel; delivered the Expedition 17 crew (Volkov and Kononenko); Yi was the first South Korean in space; Richard Garriott returned to Earth on board TMA-12. Duration was 198 days, 16 hours, 20 minutes … (Russia).

05/31/2008 … STS-124 Discovery … Mark Kelly, Kenneth Ham, Karen Nyberg, Ronald Garan, Michael Fossum, Akihiko Hoshide, Gregory Chamitoff … ISS assembly flight 1J; delivered the Japanese Kibo Pressurized Module and the Remote Manipulator System; exchanged crew members (Chamitoff for Garrett Reisman); three EVAs were conducted with two crew members, totaling 20 hours, 32 minutes. Duration was 13 days, 18 hours, 13 minutes … (USA).

09/25/2008 … Shenzhou 7 … Zhai Zhigang, Liu Boming, Jing Haipeng … The first Chinese three man-crew; the first Chinese EVA (by Zhai Zhigang, totaling 20 minutes); released a 40 kilogram sub-satellite. Duration was 2 days, 20 hours, 27 minutes … (China).

10/12/2008 … Soyuz TMA-13 … Yuri Lonchakov, Michael Fincke, Richard Garriott … Exchanged TMA-13 for TMA-12 to serve the ISS as an emergency escape vessel; delivered the Expedition 18 crew (Fincke and Lonchakov); Charles Simonyi returned to Earth on board TMA-13. Duration was 178 days, 14 minutes … (Russia).

11/15/2008 … STS-126 Endeavor … Christopher Ferguson, Eric Boe, Donald Pettit, Stephen Bowen, Sandra Magnus, Heidemarie Stefanyshyn-Piper, Robert Kimbrough … ISS utilization and logistics flight ULF-2; Leonardo MPLM; delivered the life support and habitability system; performed ISS maintenance; exchanged crew members (Magnus for Gregory Chamitoff); conducted four EVAs with three crew members, totaling 26 hours, 41 minutes. Duration was 15 days, 20 hours, 30 minutes … (USA).

03/15/2009 … STS-119 Discovery … Lee Archambault, Dominic Antonelli, Joseph Acaba, Steven Swanson, Richard Arnold, John Phillips, Koichi Wakata … ISS assembly flight 15A; installed the S6 integrated truss and the fourth set of solar arrays; exchanged crew members (Wakata for Sandra Magnus); conducted three EVAs with three crew members, totaling 19 hours, 04 minutes, Duration was 12 days, 19 hours, 30 minutes … (USA).

03/26/2009 … Soyuz TMA-14 … Gennadi Padalka, Michael Barratt, Charles Simonyi … Exchanged TMA-14 for TMA-13 to serve the ISS as an emergency escape vessel; delivered Expedition 19 and 20 crews (Padalka and Barratt); Simonyi makes the second trip into orbit as a tourist; Guy Laliberte returned to Earth on board TMA-14 on 10/11/2009. Duration was 198 days, 16 hours, 42 minutes … (Russia).

05/11/2009 ... STS-125 Atlantis ... Scott Altman, Gregory Johnson, Michael Good, Megan McArthur, John Grunsfeld, Mike Massimino, Andrew Feustel ... The fifth Hubble Space Telescope servicing mission; five EVAs were conducted with four crew members, totaling 36 hours, 56 minutes. Duration was 12 days, 21 hours, 37 minutes ... (USA).

05/27/2009 ... Soyuz TMA-15 ... Roman Romanenko, Robert Thirsk, Frank De Winne ... Delivered the Expedition 20 and 21 crew; remained docked to the ISS to serve as an emergency escape vessel; the start of six-person crew operations on the ISS. Duration was 187 days, 20 hours, 41 minutes ... (Russia).

06/15/2009 ... STS-127 Endeavor ... Mark Polansky, Douglas Hurley, Christopher Cassidy, Julie Payette, Thomas Marshburn, David Wolf, Timothy Kopra ... ISS assembly flight 2J/A; delivered the Japanese Kibo Exposed Facility and the Exposed Section of the Experiment Logistics Module; exchanged crew members (Kopra for Koichi Wakata); five EVAs were conducted with four crew members, totaling 30 hours, 30 minutes. Duration was 15 days, 16 hours, 45 minutes ... (USA).

08/29/2009 ... STS-128 Discovery ... Frederick Sturckow, Kevin Ford, Patrick Forrester, Jose Hernandez, Christer Fuglesang, John Olivas, Nicole Stott ... ISS assembly flight 17A; Leonardo MPLM; Lightweight Multi-Purpose Experiment Support Structure Carrier was delivered and attached to the ISS; exchanged crew members (Stott for Timothy Kopra); three EVAs were conducted with three crew members, totaling 20 hours, 15 minutes. Duration was 13 days, 20 hours, 54 minutes ... (USA).

09/30/2009 ... Soyuz TMA-16 ... Maksim Surayev, Jeffrey Williams, Guy Laliberte ... Exchanged TMA-16 for TMA-14 to serve the ISS as an emergency escape vessel; delivered the Expedition 21 and 22 crew (Williams and Surayev); Laliberte was the first Canadian space tourist to orbit. Duration was 169 days, 04 hours, 10 minutes ... (Russia).

11/16/2009 ... STS-129 Atlantis ... Charles Hobaugh, Barry Wilmore, Leland Melvin, Randolph Bresnik, Michael Foreman, Robert Satcher ... ISS utilization and logistics flight ULF-3; ExPRESS Logistics Carriers ELC-1 and ELC-2 were delivered; also delivered spare components; three EVAs were conducted with three crew members, totaling 18 hours, 27 minutes; Nicole Stott (STS-128) returned to Earth on board STS-129. Duration was 10 days, 19 hours, 16 minutes … (USA).

12/20/2009 … Soyuz TMA-17 … Oleg Kotov, Timothy Creamer, Soichi Noguchi … Delivered the Expedition 22 and 23 crew; remained docked to serve the ISS as an emergency escape vessel. Duration was 163 days, 05 hours, 32 minutes … (Russia).

02/08/2010 … STS-130 Endeavor … George Zamka, Terry Virts, Kathryn Hire, Stephen Robinson, Nicholas Patrick, Robert Behnken … ISS assemble flight 20A; delivered the Tranquility Module (Node 3) and the Cupola; three EVAs were conducted with two crew members, totaling 18 hours, 14 minutes. Duration was 13 days, 18 hours, 08 minutes … (USA).

04/02/2010 … Soyuz TMA-18 … Aleksandr Skvortsov, Mikhail Korniyenko, Tracy Caldwell-Dyson … Delivered the Expedition 23 and 24 crew; remained docked to the ISS to serve as an emergency escape vessel. Duration was 176 days, 01 hour, 19 minutes … (Russia).

04/05/2010 … STS-131 Discovery … Alan Poindexter, James Dutton, Richard Mastracchio, Dorothy Metcalf-Lindenburger, Stephanie Wilson, Clayton Anderson, Naoko Yamazaki … ISS assembly flight 19A; Leonardo MPLM; replaced an ammonia tank and rate gyro assemblies; three EVAs were conducted with two crew members, totaling 20 hours, 17 minutes. Duration was 15 days, 02 hours, 47 minutes … (USA).

05/14/2010 … STS-132 Atlantis … Kenneth Ham, Dominic Antonelli, Garrett Reisman, Michael Good, Stephen Bowen, Piers Sellers … ISS utilization and logistics flight ULF-4; delivered the Russian Rassvet Mini-Research Module; three EVAs were conducted with three crew members, totaling 21 hours, 20 minutes. Duration was 11 days, 18 hours, 29 minutes … (USA).

06/15/2010 … Soyuz TMA-19 … Fyodor Yurchikhin, Shannon Walker, Douglas Wheelock … Delivered Expedition 24 and 25 crews; remained docked with the ISS to serve as an emergency escape vessel. Duration was 163 days, 07 hours, 11 minutes … (Russia).

10/07/2010 ... Soyuz TMA-01M ... Aleksandr Kaleri, Oleg Skripochka, Scott Kelly ... Delivered the Expedition 25 and 26 crews; remained docked with the ISS to serve as an emergency escape vessel. Duration was 159 days, 08 hours, 43 minutes ... (Russia).

12/15/2010 ... Soyuz TMA-20 ... Dmitri Kondratiyev, Paolo Nespoli, Catherine Coleman ... Delivered the Expedition 26 and 27 crews; remained docked with the ISS to serve as an emergency escape vessel. Duration was 159 days, 07 hours, 18 minutes ... (Russia).

02/24/2011 ... STS-133 Discovery ... Steven Lindsey, Eric Boe, Nicole Stott, Alvin Drew, Michael Barratt, Stephen Bowen ... ISS utilization and logistics flight ULF-5; ExPRESS Logistics Carrier ELC-4 was delivered, as well as the Permanent Multi-Purpose Module (PMM); two EVAs were conducted with two crew members, totaling 12 hours, 48 minutes; this was the last flight for the Discovery. Duration was 12 days, 19 hours, 05 minutes ... (USA).

04/04/2011 ... Soyuz TMA-21 ... Andrei Borisenko, Aleksandr Samokutyayev, Ronald Garan ... Delivered the Expedition 27 and 28 crews; remained docked with the ISS to serve as an emergency escape vessel. Duration was 164 days, 18 hours, 58 minutes ... (Russia).

05/16/2011 ... STS-134 Endeavor ... Mark Kelly, Gregory Johnson, Michael Fincke, Roberto Vittori, Andrew Feustel, Gregory Chamitoff ... ISS utilization and logistics flight ULF-6; ExPRESS Logistics Carrier ELC-3; Alpha Magnetic Spectrometer (AMS); four EVAs were conducted with three crew members, totaling 28 hours, 44 minutes; this was the last flight for the Endeavor. Duration was 15 days, 17 hours, 39 minutes ... (USA).

06/07/2011 ... Soyuz TMA-02M ... Sergei Volkov, Michael Fossum, Satoshi Furukawa ... Delivered the Expedition 28 and 29 crew; remained docked with the ISS to serve as an emergency escape vessel. Duration was 167 days, 04 hours, 14 minutes ... (Russia).

07/08/2011 ... STS-135 Atlantis ... Christopher Ferguson, Douglas Hurley, Sandra Magnus, Rex Walheim ... ISS utilization and logistics flight ULF-7; Raffaello MPLM; Lightweight Multi-Purpose Carrier (LMC); this was the last flight for the Atlantis, and the end of the American STS shuttle program. Duration was 12 days, 18 hours, 29 minutes … (USA).

11/14/2011 … Soyuz TMA-22 … Anton Shkaplerov, Anatoli Ivanishin, Daniel Burbank ... Combined operations with the Expedition 29 crew; remained docked with the ISS to serve as an emergency escape vessel. Duration was 165 days, 07 hours, 31 minutes … (Russia).

12/21/2011 … Soyuz TMA-03M … Oleg Kononenko, Andre Kuipers, Donald Petit ... Combined operations with the Expedition 30 crew; remained docked to the ISS to serve as an emergency escape vessel. Duration was 192 days, 18 hours, 58 minutes … (Russia).

05/15/2012 … Soyuz TMA-04M … Gennady Padalka, Sergei Revin, Joseph Acaba ... Combined operations with the Expedition 31 crew; remained docked with the ISS to serve as an emergency escape vessel. Duration was 124 days, 23 hours, 52 minutes … (Russia).

06/16/2012 … Shenzhou 9 … Jing Haipeng, Liu Wang, Liu Yang ... Docked with the unmanned Tiangong-1 experimental laboratory; Shenzhou 8 (launched on 10/31/2011) was reported by the Chinese to be an unmanned test of automatic docking techniques with the Tiangong-1; Yang was the first Chinese woman in space. Duration was 12 days, 03 hours, 24 minutes … (China).

07/15/2012 … Soyuz TMA-05M … Yuri Malenchenko, Sunita Williams, Akihiko Hoshide ... Combined operations with the Expedition 32 crew; remained docked with the ISS to serve as an emergency escape vessel. Duration was 126 days, 11 hours, 16 minutes … (Russia).

10/23/2012 … Soyuz TMA-06M … Oleg Novitski, Evgeny Tarelkin, Kevin Ford ... Combined operations with the Expedition 33 crew; remained docked with the ISS to serve as an emergency escape vessel. The crew reentered Soyuz TMA-06M and undocked from the ISS at 1943 hours (EST) on 03/15/2013 for re-entry. They landed safely in Russia at precisely 0306 hours (GMT) on Saturday, 03/16/2013 … Duration was 144 days, 11 hours, 49 minutes … (Russia).

12/19/2012 ... Soyuz TMA-07M ... Roman Romanenko, Thomas Marshburn, Chris Hadfield ... Expedition 34 crew; remained docked with the ISS to serve as an emergency escape vessel. The Canadian astronaut Chris Hadfield became the Expedition 35 commander when Soyuz TMA-06M undocked from the ISS on 03/15/2013 ... (Russia).

03/28/2013 ... Soyuz TMA-08M ... Pavel Vinogradov, Aleksandr Misurkin, Christopher Cassidy ... Expedition 35 crew; the first quick (six hours) rendezvous to the ISS. Mission duration was 166 days, 6 hours, 15 minutes. Undocked from the ISS at 2337 GMT on 09/10/2013. Landed on 09/11/2013 at 0258 UTC ... (Russia).

05/28/2013 ... Soyuz TMA-09M ... Fyodor Yurchikhin (of the RSA), Karen L. Nyberg (of NASA), Luca Parmitano (of the ESA) ... Expedition 36 crew; remained docked to the ISS during Expeditions 36 and 37 to serve as an emergency escape vessel. Mission duration was 166 days, 6 hours, 18 minutes. Undocked from the ISS at 2326 UTC on 11/10/2013. Landed on 11/11/2013 at 0349 UTC ... (Russia).

06/11/2013 ... Shenzhou 10 ... Nie Haisheng, Zhang Xiaoguang, Wang Yaping ... Docked with the Tiangong-1 trial space laboratory on 06/13/2013. Mission duration was 14 days, 14 hours, 29 minutes. Undocked on 06/25/2013. Landed on 06/26/2013 at 00:07 UTC in inner Mongolia ... (China).

09/25/2013 ... Soyuz TMA-10M ... Oleg Kotov (RSA), Sergey Ryazansky (RSA), Michael Hopkins (NASA) ... Expedition 37 crew members; remained docked to the ISS during the Expedition 38 and 39 increment to serve as an emergency escape vessel. Mission duration was 5 months, 13 days, 6 hours, 26 minutes. Launch was at 20:59:50 UTC on 09/25/2013. Docking with the ISS was at 0245 UTC on 09/26/2013. Undocking was at 0002 UTC on 03/11/2014. Landing was at 03:24 UTC on 03/11/2014 ... (Russia).

11/07/2013 ... Soyuz TMA-11M (call sign Vostok) ... Mikhail Tyurin (RSA), Richard Mastracchio (NASA), Koichi Wakata (JAXA) ... Expedition 38 crew members Launch was at 04:14:15 UTC on 11/07/2013. Docked with the ISS on 11/07/2013 at 1027 UTC. The flight carried the

2014 Winter Olympic torch … it was returned to Earth five days later on board Soyuz TMA-09M … (Russia).

03/25/2014 … Soyuz TMA-12M … (call sign Cliff) … Aleksandr Skvortsov (RSA), Oleg Artemyev (RSA), Steven R. Swanson (NASA) … Expedition 39 crew members to the ISS. Launch was at 2117 UTC on 03/25/2014. Docking with the ISS was at 2353 UTC on 03/27/2014 … it was returned to Earth on 09/11/2014 at 02:23 UTC … (Russia).

05/28/2014 … Soyuz TMA-13M … (call sign Cepheus) … Maksim Surayev (RSA), Gregory R. Wiseman (NASA), Alexander Gerst (ESA) … Expedition 40 crew members to the ISS. Launch was at 19:57:41 UTC on 05/28/2014. Docking with the ISS was at 01:44 on 05/29/2014 … the return to Earth date was on 11/10/2014 at precisely 03:58 UTC … (Russia).

09/25/2014 … Soyuz TMA-14M … Aleksandr Samokutyayev (RSA), Yelena Serova (RSA), Barry E. Wilmore (NASA) … Expedition 41 crew members to the ISS. Yelena Serova is the first female Russian cosmonaut to live and work on board the ISS. The port-side solar array panel failed to deploy, hence the vessel operated on its reserve power system and the electricity generated by the starboard-side solar array panel that did deploy correctly. The launch was at 20:25:00 UTC on 09/25/2014, and the return to Earth was on 03/12/2015 at 02:07 UTC … (Russia).

10/31/2014 … Virgin Galactic SpaceshipTwo … Michael Alsbury, Peter Siebold ... The fourth powered test flight of the SpaceShipTwo suborbital craft designed to fly up to six people into space approximately 60 miles above the Earth. There were no passengers on this test flight, just the two astronauts. The craft separated successfully from the carrier aircraft (White Knight Two) at around 50,000 feet, and, for the first few seconds all seemed well. Upon the ignition of SpaceShipTwo's rocket motor, however, there was an apparent explosion that destroyed the spacecraft. Peter Siebold managed to eject from the spacecraft, and landed with severe injuries. He was taken to a local area hospital in California. Michael Alsbury was killed in the explosion, his body found later in the wreckage. A timeline of events leading up to the catastrophic breakup was stated by the NTSB chairman as follows: 10:07:19 SpaceShipTwo is released from the

carrier craft WhteKnightTwo. 10:07:21 SpaceShipTwo's engine starts. 10:07:29 SpaceShipTwo reaches mach 0.94, nearly the speed of sound. 10:07:31 SpaceShipeTwo exceeds the speed of sound at mach 1.02. Between 10:07:29 and 10:07:31 a rudder feathering safety device was unlocked at around mach 1.02. The feathering device was designed to be unlocked at Mach 1.04, however. The incident is under investigation by the NTSB, the National Transportation Safety Board. The launch occurred on 10/31/2014 ... (USA).

11/23/2014 ... Soyuz TMA-15M ... call sign for this mission was "Astraeus" ... Anton Shkaplerov (RSA), Samantha Cristoforetti (ESA), Terry W. Virts (NASA) ... Expedition 42 crew members to the International Space Station (ISS). The launch was at 21:01 UTC on 11/23/2014, with a planned return to Earth time frame in May of 2015. It remained docked to the ISS for the Expedition 43 increment to serve as an emergency escape vessel. The actual return to Earth was on 06/11/2015 at 13:44 UTC ... (Russia).

03/27/2015 ... Soyuz TMA-16M ... Gennady Padalka (RSA), Mikhail Korniyenko (RSA), Scott Kelly (NASA) ... Padalka and Korniycnko returned on 09/12/2015, and Kelly returned on Soyuz TMA-18M after a full year in orbit ... (Russia).

07/22/2015 ... Soyuz TMA-17M ... Oleg Kononenko (RSA), Kimiya Yui (JAXA), Kjell N. Lindgren (NASA) ... this was the Expedition 44 mission to the ISS ... the return to Earth was on 12/11/2015 ... (Russia).

09/02/2015 ... Soyuz TMA-18M ... Sergey Volkov (RSA), Andreas Mogensen (ESA), Aidyn Aimbetov (KazCosmos) ... this was the Expedition 45 mission to the ISS ... the return to Earth was on 03/02/2016 ... (Russia).

12/15/2015 ... Soyuz TMA-19M ... Yuri Malenchenko (RSA), Timothy Kopra (NASA), Timothy Peake (ESA) ... this was the Expedition 46 mission to the ISS ... the return to Earth was on 06/18/2016 ... (Russia).

03/18/2016 ... Soyuz TMA-20M ... call sign was "Burlak" ... Aleksey Ovchinin (RSA), Oleg Skripochka (RSA), Jeffrey Williams (NASA) ... this was the Expedition 47 mission to the ISS … the return to Earth was on 09/07/2016 … (Russia).

07/07/2016 … Soyuz MS-01 … the call sign was "Irkut" … Anatoli Ivanishin (RSA), Takuya Onishi (JAXA), Kathleen Rubins (NASA) … this was the Expedition 48 flight to the ISS … return to Earth was on 10/30/2016 … (Russia).

10/17/2016 … Shenzou 11 … Jing Haipeng, Chen Dong … launched from the Jiuquan Satellite Launch Center using a Long March 2F rocket, it docked with the Chinese Tiangong-2 space laboratory … the return to Earth was 33 days later on 11/18/2016 … (China).

10/19/2016 … Soyuz MS-02 … Sergey Nikolayevich (RSA), Andrei Borisenko (RSA), Robert S. Kimbrough (NASA) … this was the Expedition 49 mission to the ISS … return to Earth was on 04/10/2017 … (Russia).

11/17/2016 … Soyuz MS-03 … Oleg Novitskiy (RSA), Thomas Pesquet (ESA), Peggy A. Whitson (NASA) … this was the Expedition 50 mission to the ISS … Peggy Whitson, at age 56, became the oldest woman to fly into space … the return to Earth was on 06/02/2017 … (Russia).

04/20/2017 … Soyuz MS-04 … the call sign was "Altair" … Fyodor Yurchikhin (RSA), Jack D. Fischer (NASA) … this was the Expedition 51 mission to the ISS … the return to Earth was on 09/03/2017 … (Russia).

07/28/2017 … Soyuz MS-05 … Sergey Ryazansky (RSA), Paolo Nespoli (ESA), Randy Bresnik (NASA) … this was the Expedition 52 flight to the ISS … this was also the 134th flight of the Soyuz spacecraft design … the return to Earth was on 12/14/2017 … (Russia).

09/12/2017 … Soyuz MS-06 … Alexander Misurkin (RSA), Mark Vande Hei (NASA), Joseph Acaba (NASA) … this was the Expedition 53 mission to the ISS … the return to Earth was on 02/28/2018 … (Russia).

12/17/2017 … Soyuz MS-07 … Aleksander Skvortsov (RSA), Norishige Kanai (JAXA), Scott D. Tingle (NASA) … this was the

Expedition 54 flight to the ISS … the return to Earth was on 06/03/2018 … (Russia).

03/21/2018 … Soyuz MS-08 … Oleg Artemyev (RSA), Andrew J. Feustel (NASA), Richard R. Arnold (NASA) … this was the Expedition 55 flight to the ISS … the return to Earth was on 10/04/2018 … (Russia).

06/06/2018 … Soyuz MS-09 … Sergey Prokopyev (RSA), Serena M. Aunon-Chancellor (NASA), Alexander Gerst (ESA) … this was the Expedition 56 and 57 flight to the ISS … with a mission that was anything but mundane, its three person crew returned to Earth after more than six months in orbit on the ISS. The Soyuz MS-09 became a high profile mission of extreme urgency after it was discovered that a small hole had been purposefully drilled into the side of its orbital module. That hole was determined to have been drilled into the spacecraft by someone on the ground prior to the flight, and during its manufacturing. It led to a dramatic spacewalk early in December of 2018, that saw cosmonauts cut into the protective casing, in order to conduct further investigations. And then there was a full blown investigation by Russian authorities who were trying to determine who was responsible for what could have been something nothing short of a tragedy, a virtual space related homicide of three people. But the spacecraft managed to land safely, after undocking from the ISS at 20:42 EST, (01:42 UTC), on Thursday, 12/20/2018, and landing at 00:03 EST (05:03 UTC) on 12/20/2018 … (Russia).

10/11/2018 … Soyuz MS-10 … Dmitry Rogozin (RSA), Nick Hague (NASA) … this was a launch failure with an emergency landing, with the two crew members experiencing high G forces due to the violent and extreme angle by which the spacecraft dropped after the failed second stage ignition. The landing in Siberia was a bone-jarring event, and the two man crew was extremely fortunate not to have suffered serious injuries or death. The Russians, not unlike the former Soviet mindset, have been far less than forthcoming about the cause of this failure, but it was certain to have set the Russian human space program back for several months … (Russia).

12/03/2018 … Soyuz MS-11 … Oleg Kononenko (RSA), David Saint-Jaques (CSA), Ann McClain (NASA) … flight to the ISS … Soyuz MS-11 landed in Dzheskasgan in Russia on 06/25/2019 with McClain, Sant-Jaques and Kononenko on board … (Russia).

12/13/2018 … Virgin Galatic VSS Unity … Mark Stucky and Frederick W. Sturckolo … a successful sub-orbital flight up to an altitude of 51.4 miles … (USA).

02/22/2019 … Virgin Galatic VSS Unity … Beth Moses, Dave Mackay and Michael Masucci … the second successful sub-orbital flight up to an altitude of 55.85 miles … the first time this spacecraft had three people on board … (USA).

03/14/2019 … Soyuz MS-12 … Aleksey Ovchinin (RSA), Christina Koch (NASA), Nick Hague (NASA) … Expedition 59 flight to the ISS … returned to Earth on 10/03/2019 … (Russia).

07/20/2019 … Soyuz MS-13 … Aleksandr Skvortsov (RSA), Luca Parmitano (Italy), Andrew R. Morgan (NASA) ...Expedition 60 flight to the ISS … returned to Earth on 02/06/2020 with Christina Koch (who undertook six EVAs and broke all previous records for the longest time in space by any woman), leaving Andrew R. Morgan aboard the ISS … (Russia).

09/25/2019 … Soyuz MS-15 … Oleg Skripochka (RSA), Jessica Meir (NASA), Hazza Al Mansouri (UAE) … Expedition 61 flight to the ISS … returned to Earth on 04/17/2020, carrying Andrew R. Morgan (NASA) … (Russia).

04/09/2020 … Soyuz MS-16 … Anatoli Ivanishin (RSA), Ivan Vagner (RSA), Chistopher Cassidy (NASA) … Expedition 62 flight to the ISS … after 195 days in orbit, the crew returned to Earth on 10/22/2020 at 02:54:12 UTC … (Russia).

05/30/2020 … (at 19:22:45 UTC) … SpaceX's Crew Dragon (Demo 2 … Crew Dragon Capsule 206, named Endeavor) … Doug Hurley (NASA), Bob Behnken (NASA) … the first crewed flight of the SpaceX series of spacecraft to the ISS, and the first all-American space flight launched from Cape Canaveral since the naïve, irresponsible, and extremely unfortunate early cancellation of the Shuttle program in 2011. They named their Dragon capsule "Endeavor" because both men had previously flown on the shuttle Endeavor … it was launched on top of a Falcon 9 rocket ... (finally, American astronauts again launched on-board American spacecraft, and from American soil!) … returned to Earth with both astronauts on August the 2nd, 2020, at precisely 18:48:06 in the Gulf of Mexico off the coast of Pensacola, Florida … the next launch with the same reusable capsule is tentatively planned for the spring of 2021 … returned to Earth on 08/02/2020 ...(USA).

10/14/2020 … Soyuz MS-17 … Sergey Ryzhikov (RSA), Sergey Kud-Sverchkov (RSA), Kathleen Rubins (NASA) … Expedition 63 flight to the ISS … returned to Earth on 04/17/2021 … (Russia).

11/15/2020 … SpaceX's Crew-1 mission, named Resilience … Michael Hopkins (NASA), Victor Glover (NASA), Shannon Walker (NASA), Soichi Noguchi (Japan) … flight to the ISS … returned to Earth on 05/02/2021 … (USA).

04/09/2021 … Soyuz MS-18 … Oleg Novitsky (RSA), Pyotr Dubrov (RSA), Mark T. Vande Hei (NASA) … flight to the ISS … the return to Earth was on 10/17/2021 at 12:35 EST … it carried Oleg Novitsky, as well as both Yulia Peresild and Klim Shipenko who flew to the ISS on the Soyuz MS-19 on 10/05/2021 to attempt a movie production entitled "Challenge" … (Russia).

04/23/2021 … SpaceX's Crew-2 mission … K. Megan McArthur (NASA), R. Shane Kimbrough (NASA), Thomas Pesquet (ESA … France), Akihiko Hoshide (Japan) … flight to the ISS … the return to Earth was on 11/08/2021 at precisely 10:33p.m. EST in the Atlantic Ocean … (USA).

05/22/2021 … Virgin Galactic VSS Unity 21 spacecraft … Frederick C.J. Sturckow and Dave MacKay (both test pilots/astronauts with Virgin Galactic/NASA) … sub-orbital flight out of the New Mexico's spaceport, south-east of Truth or Consequences … the flight reach an altitude of 55.45 miles, taking off at 8:35a.m. MDT (1435 GMT), and landing at the same spaceport in New Mexico at 9:43a.m. (1543 GMT) … (USA).

06/17/2021 … Shenzhou-12 … Nie Hasheng, Liu Boming, Tang Hongbo … this was the first human occupied space flight to the Chinese space station, the Tianhe module, and the crew is supposed to spend approximately three months in orbit … the return to Earth was on 09/17/2021 … (China).

07/11/22021 … Virgin Galactic VSS Unity 22 spacecraft … the pilots were David MacKay and Michael Masucci, and the four passengers were Sirisha Bandla, Colin Bennett, Beth Moses, and Richard Branson … this was the first passenger sub-orbital flight for Virgin Galactic … it reached an altitude of 53.551 miles, before landing back at the New Mexico spaceport at 15:40 UTC … (USA).

07/20/2021 … Blue Origin's New Shepard spacecraft … this was the first human occupied flight of the Blue Origin spacecraft … it was automatically piloted from the ground … the four passengers were Jeff Bezos, Wally Funk (at 82 years of age, she was the oldest person to venture into space), Oliver Daemon (at 18 years of age, he was the youngest person to venture into space), and Mark Bezos … the launch was from the company's spaceport located in Van Horn, Texas … it reached a sub-orbital altitude of 347,563 feet, approximately 65 miles, and the flight lasted for 10 minutes and 10 seconds … (USA).

09/15/2021 … SpaceX's Inspiration4 mission … Jared Isaacman, Hayley Arceneaux, Sian Proctor, Christoper Sembroski … an all civilian crew … they orbited the Earth for three days at an approximate altitude of 364 miles, while conducting various experiments, before returning to a splashdown in the Atlantic Ocean … the launch time was 00:02:56 UTC, and the landing time was 23:06:49 UTC on 09/18/2021 … (USA).

10/05/2021 … Soyuz MS-19 … Anton Shkaplerov (RSA), Yulia Peresild (an actress), and Kilm Shipenko (a film producer) … launch to the International Space Station, where they will film some of the activities for a movie production … launch time was 4:55 EDT … the landing date was 03/30/2022 at 11:28:26 UTC, 147 kilometers southeast of Zhezkazgan, and brought back to Earth Mark T. Vande Hei of NASA, as well as both Pyotr Dubron and Anton Shkaplerov of Roscosmos … (Russia).

10/13/2021 … Blue Origin's New Shepard NS-18 sub-orbital flight … William Shatner (Captain Kirk of "Star Trek" television fame), Audrey Powers, Chris Boshuizen, Glen de Vries … it reached a brief altitude of 106 kilometers (65.8653 miles) … (USA).

11/11/2021 … SpaceX's Crew 3 Mission to the ISS … NASA astronauts Tom Marshburn, Kayla Barron and Raja Chari, as well as European Space Agency astronaut Matthias Maurer … the launch was conducted precisely at 02:03:31 UTC … it is scheduled for a six month stay on-board the ISS … (USA).

12/08/2021 … Soyuz MS-20 … Alexander Misurkin (RSA), and two space tourists, Yusaku Maezawa and Yozo Hirano, both from Japan … they flew to the ISS for an twelve day stay, and landed back on Earth on 12/20/2021 at 03:13 UTC … (Russia).

12/11/2021 … Blue Origin New Shepard NS-19 sub-orbital spaceflight … Dylan Taylor, Lane Bess, Cameron Bess, Laura Shepard Churchley, Michael Stahan, Evan Dick … an all civilian crew … launch was at 15:00:42 UTC and reached an altitude of 62 miles … the launch and landing were in Texas, with a duration of 10 minutes … Laura Shepard Churchley is the daughter of the late astronaut Alan Shepard … (USA).

03/31/2022 … Blue Origin New Shepard NS-20 sub-orbital spaceflight … Marty Allen, Sharon Hagle, Marc Hagle, Jim Kitchen, Gary Lai, and Dr. George Nield … an all civilian crew … launch was at 13:58 UTC and reached an altitude of 347,629 feet, approximately 65.7 miles … the launch and landing were in west Texas, with a duration of 10 minutes, 4 seconds … (USA).

04/08/2022 … SpaceX Axiom 1 flight (Ax-1) … Michael LopezAlegria, Larry Connor, Mark Pathy; Eytan Stibbe … an all civilian crew … launch was at 15:17:52 UTC … the spacecraft docked with the ISS at 8:29a.m. EDT on 04/09/2022 … the flight to the ISS was for a ten day stay, with a return scheduled for 04/18/2022 … (USA).

04/27/2022 ... SpaceX Crew 4 ... Kjell N. Lindgren (NASA), Robert Hines (NASA), Samantha Cristoforetti (ESA), Jessica Watkins (NASA) ... flight to the ISS ... landing date was 10/14/2022 ... (USA).

06/04/2022 ... Blue Origin NS-21 ... Evan Dick, Katya Echazarreta, Hamish Harding, Victor Correa Hespanha, Jaison Robinson, Victor Vescovo ... a sub-oribtal flight lasting 10 minutes and 5 seconds ... landing date was 06/04/2022 ... (USA).

06/05/2022 ... Shenzhou 14 ... Chen Dong, Liu Yang, Cai Xuzhe ... flight to the Chinese space station ... landing date was 12/04/2022 ... (China).

08/04/2022 ... Blue Origin NS-22 ... Coby Cotton, Mario Ferreira, Vanessa O'Brien, Clint Kelly III, Sara Sabry, Steve Young ... a suborbital flight lasting 10 minutes and 20 seconds ... landing date was 08/04/2022 ... (USA).

09/21/2022 ... Soyuz MS-22 ... Sergey Prokopyev, Dmitry Petelin, Francisco Rubio (NASA) ... flight to the ISS ... landing date was 03/28/2023 ... (Russia).

10/05/2022 ... SpaceX Crew 5 ... Nicole Aunapu Mann (NASA), Josh A. Cassada (NASA), Koichi Wakata (JAXA), Anna Kikina (Roscosmos) ... flight to the ISS ... landing date was 03/12/2023 ... (USA).

11/29/2022 ... Shenzhou 15 ... Fei Junlong, Deng Qingming, Zhang Lu ... flight to the Chinese space station ... landing date was 06/03/2023 ... (China).

03/02/2023 ... SpaceX Crew 6 ... Stephen Bowen (NASA), Warren Hoburg (NASA), Sultan Al Neyadi (MBRSC), Andrey Fedyaev (Roscosmos) ... flight to the ISS ... (USA).

05/21/2023 ... Axiom Mission 2 ... Peggy Whitson (Axiom Space), John Shoffner (commercial astronaut), Ali AlQami (SSC), Rayyanah Barnawi (SSC) ... flight to the ISS ... landing date was 05/31/2023 ... (USA).

05/25/2023 ... Virgin Galactic Unity 25 ... Michael Masucci, Frederick W. "CJ" Sturckow, Beth Moses, Luke Mays, Jamila Gilbert, Christopher Huie ... a sub-orbiotal flight ... (USA).

05/30/2023 ... Shenzhou 16 ... Jing Haipeng, Zhu Yangzhu, Gui Haichao ... flight to the Chinese space station ... the landing was on 10/31/2023 at 00:12 UTC in inner Mongolia, China ... (China).

06/29/2023 ... Galactic 01 ... Michael Masucci, Nicola Pecile, Walter Villader, Pantal Carlucci, Angelo Landolfi, Colin Bennett ... previously referred to as Unity 23, this was another sub-orbital spaceflight of the StarShipTwo class VSS Unity ... it was the first commercial spaceflight for Virgin Galactic ... launch was at 14:30:00 UTC, and landing was at 15:42:28 UTC ... (USA).

08/10/2023 ... Galactic 02 ... CJ Sturckow, Kelly Latimer, Beth Moses, Jon Goodwin, Keisha Schahaff, Anastatia Mayers ... the second sub-orbital commercial spaceflight of the StarShipTwo class VSS Unity ... launch was at 14:29:45 UTC, and reached an alititude of 55.0 miles ... (USA).

08/26/2023 ... SpaceX Crew 7 ... Jasmin Moghbeli (NASA), Andreas Mogensen (ESA), Satoshi Furukawa (JAXA), Konstantin Borisov (Roscosmos) ... launch was at 07:27:27 UTC to the ISS ... the planned landing date is in the first quarter of 2024 ... (USA).

09/08/2023 ... Galactic 03 ... Nicola Pecile, Michael Masucci, Beth Moses, Ken Baxter, Timothy Nash, Adrian Reynard ... launch was at 14:34 UTC ... a sub-orbital flight that reach an altitude of 55.03 miles, and lasted until 15:36 UTC ... (USA).

09/15/2023 ... Soyuz MS-24 ... Oleg Kononenko (Roscosmos), Nikolai Chub (Roscosmos), Loral O'Hara (NASA) ... launch was at 15:44 UTC to the ISS ... (Russia).

10/06/2023 ... Galactic 04 ... Kelly Latimer, CJ Sturchow, Beth Moses, Ron Rosano, Trevor Beattie, Namira Salim ... launch was at 15:28 UTC ... a sub-orbital flight that reached an altitude of 54.03 miles, and lasted until 16:23 UTC ... (USA).

10/26/2023 ... Shenzhou 17 ... Tang Honbo, Tang Shengjie, Jiang Xinlin ... flight to the Chinese space station ... (China).

11/02/2023 ... Galactic 05 ... Michael Masucci, Kelly Latimer, Colin Bennett, Alan Stern, Kellie Gerardi, Ketty Maisonrouge ... launch was at 15:00 UTC ... a sub-orbital flight that reached an altitude of 54.2 miles, and lasted until 15:59 UTC ... (USA).

01/18/2024 ... Axiom Mission 3 ... Michael Lopez-Alegria, Walter Villadi, Alper Gezeravci, Marcus Wandt ... launch was at 21:49 UTC ... flight to the ISS ... (USA).

01/26/2024 ... Galatic 06 ... CJ Sturckow, Nicola Pecile, Lina Borozdina, Robie Vaughn, Franz Haider, Neil Kornswiet ... launch was at 17:00 UTC ... a sub-orbital flight that reached at altitude of 55.2 miles, and lasted until 17:56 UTC ... (USA).

———

There will be many more human-occupied space flights, and, hopefully, we here in the United States, in addition to the new SpaceX series of spacecraft that are in their infancy, will soon see the launch of the Orion and other American spaceship designs carrying us back to the Moon and then onwards to Mars. And, with thanks to President Donald Trump's prior leadership, getting back to the Moon, and then going on to Mars, just may become a reality.

———

For the 2027 seventy-year anniversary of the first human space flight, there will be a second volume of this extraordinary expose!

Abbreviations

ASTP … The Apollo Soyuz Test Program.

Astronaut … The term used to describe Americans who go into space. See also Russian cosmonauts and Chinese taikonauts.

ATLAS-1 … Atmospheric Laboratory for Applications and Science; STS-45; USA.

ATLAS-2 … Atmospheric Laboratory; STS-56; USA.

ATLAS-3 … Atmospheric Laboratory; STS-66; USA.

Atlantis … The Space Shuttle Atlantis (Orbiter Vehicle Designation: OV-104) was a Space Shuttle orbiter in the Space Shuttle fleet belonging to NASA. Atlantis was the fourth operational (and the next-to-the-last) Space Shuttle to be constructed by the Rockwell International Company in southern California, and it was delivered to the John F. Kennedy Space Center in eastern Florida in April of 1985. Atlantis was the only orbiter which lacked the ability to draw power from the International Space Station while docked there; it had to continue to provide its own power through fuel cells. The last mission of Atlantis was STS-135, the last flight of the Shuttle program. This final flight, authorized in October of 2010, brought additional supplies to the International Space Station and took advantage of the processing performed for the Launch on the Need mission, which would only have been flown in the event that Endeavor's STS-134 crew required rescue. Atlantis launched successfully for the final time on July the 8th, 2011 at 16:29 UTC, landing at the John F. Kennedy Space Center on the July the 21st, 2011 at 09:57 UTC. By the end of its final mission, Atlantis had orbited the Earth 4,848 times, traveling nearly 126,000,000 miles (203,000,000 km) in space or more than 525 times the distance from the Earth to the Moon. Atlantis was named after RV Atlantis, a two-masted sailing ship that operated as the primary research vessel for the Woods Hole Oceanographic Institution from 1930 to 1966.

Aurora 7 ... Mercury-Atlas 7 was the second American orbital Mercury program manned space mission, launched on May the 24th, 1962. The Mercury spacecraft was named Aurora 7 and made three Earth orbits, piloted by astronaut Scott Carpenter. A targeting mishap during reentry took the spacecraft 250 miles (about 400 km) off course, delaying recovery of Carpenter and the craft. The mission used Mercury spacecraft number 18 and Atlas launch vehicle number 107-D.

CM ... Command Module for the Apollo program.

Challenger ... The Space Shuttle Challenger (NASA Orbiter Vehicle Designation OV-099) was NASA's second Space Shuttle orbiter to be put into service, Columbia having been the first. The shuttle was built by Rockwell International's Space Transportation Systems Division in Downey, California. Its maiden flight was on April the 4th, 1983, and it completed nine missions before breaking apart 73 seconds after the launch of its tenth mission, STS-51-L on January the 28th, 1986, resulting in the death of all seven crew members. It was the first of two shuttles (the other being Columbia) to be destroyed. The accident led to a two-and-a-half year grounding of the shuttle fleet, with missions resuming in 1988 with the launch of Space Shuttle Discovery on STS-26. Challenger itself was replaced by the Space Shuttle Endeavor, which first launched in May of 1992 and was constructed from structural spares that had been ordered by NASA as part of the construction contracts for Discovery and Atlantis.

Clark, Phillip S. ... A well known authority on the early Soviet and Chinese human space flight programs, and an author on the subject of space exploration. He passed away on January the 14th, 2021.

CNSA ... The China National Space Agency.

Columbia ... The Space Shuttle Columbia (NASA Orbital Vehicle Designation: OV-102) was the first spaceworthy space shuttle in NASA's orbital fleet. First launched on the STS-1 mission, the first of the Space Shuttle program, it completed 27 missions before disintegrating during re-entry on February the 1st, 2003 near the end of its 28th mission, STS-107, resulting in the deaths of all crew members aboard.

CONCAP-IV ... STS-59; USA.

Cosmonaut ... A Russian astronaut.

Cosmos ... The Cosmos series was a project by Cosmos Studios and The Planetary Society to test a solar sail in space. As part of the project, an unmanned solar sail spacecraft christened Cosmos 1 was launched into space at 15:46:09 EDT (19:46:09 UTC) on June the 21st, 2005 from the submarine Borisoglebsk in the Barents Sea. However, a rocket failure prevented the spacecraft from reaching its intended orbit. Once in orbit, the spacecraft was supposed to deploy a large sail, upon which photons from the Sun would push, thereby increasing the spacecraft's velocity (the contributions from the solar wind are similar, but of much smaller magnitude).

CRISTA/SPAS ... STS-66; USA.

D1 ... German Spacelab mission; USA.

D2 ... German Spacelab mission; STS-55; USA.

DEE ... STS-62; USA.

Discovery ... The Space Shuttle Discovery (Orbiter Vehicle Designation OV-103) is one of the retired orbiters of the Space Shuttle program of NASA, and was operational from its maiden flight, STS-41-D on August the 30th, 1984, until its final landing during STS-133 on March the 9th, 2011. Discovery has flown more than any other spacecraft having completed 39 successful missions in over 27 years of service. In 1984, Discovery became the third operational orbiter following Columbia and Challenger, and made its final touchdown at Kennedy Space Center on March the 9th, 2011 at 10:57:17 CST, having spent a cumulative total of almost a full year in space. Discovery has performed both research and International Space Station (ISS) assembly missions. Discovery also flew the Hubble Space Telescope into orbit. Discovery was the first operational shuttle to be retired, followed by Endeavor and then Atlantis.

DOD … Department of Defense; USA.

DXS … Differential X-ray Experiment: STS-54; USA.

EDO … Extended Duration Orbiter; STS-50; STS-58; USA.

Endeavor … The Space Shuttle Endeavor (Orbiter Vehicle Designation: OV-105) is one of the retired orbiters of the Space Shuttle program of NASA. Endeavor was the fifth and final spaceworthy NASA space shuttle to be built, and first flew in May of 1992 on mission STS-49 and its last mission STS-134 was in May of 2011. The STS-134 mission was originally planned as the final mission of the Space Shuttle program, but with authorization of the STS-135 mission, Atlantis became the last Space Shuttle to fly.

The United States Congress authorized the construction of Endeavour in 1987 to replace Challenger, which was lost in the STS-51-L launch accident in 1986. Structural spares built during the construction of Discovery and Atlantis, two of the previous shuttles, were used in its assembly. NASA chose to build Endeavour from spares rather than refitting Enterprise or accepting a Rockwell International proposal to build two shuttles for the price of one of the original shuttles, on the basis of cost considerations.

ESA … The European Space Agency.

EVA … Extra Vehicular Activity.

FAI … At the start of the 20th Century, the pioneering flights of pilots such as Clement Ader, the Wright Brothers and Santos-Dumont, the proliferation of aeronautical competitions, and increasingly rapid technological advances marked the real birth of the modern aviation era.

A small group of men recognized the growing need for an international federation to coordinate and give direction to the rapidly growing aeronautical activity. On the 10th of June, 1905, Count Henri de la Vaulx, Vice President of the Aero Club of France, Major Moedebeck of the German Airship League and Fernand Jacobs, President of the Aero Club of Belgium, gave a presentation to the Olympic Congress of Brussels on their proposal for a "Fédération

Aéronautique Internationale". The delegates received the idea warmly, and in token of its support the Olympic Congress adopted the following resolution: "This Congress, recognizing the special importance of aeronautics, expresses the desire that in each country, there be created an Association for regulating the sport of flying and that thereafter there be formed a Universal Aeronautical Federation to regulate the various aviation meetings and advance the science and sport of Aeronautics."

Faith 7 … Mercury-Atlas 9 was the final manned space mission of the U.S. Mercury program, launched on May the 15th, 1963 from Launch Complex 14 at Cape Canaveral, Florida. The spacecraft, named Faith 7, completed 22 Earth orbits before splashing down in the Pacific Ocean, piloted by astronaut Gordon Cooper, then an Air Force major. The Atlas rocket was number 130-D, and the Mercury spacecraft was number 20.

Freedom 7 … The Mercury-Redstone 3, MR-3 or Freedom 7 spaceflight was the first human spaceflight by the USA and took place on May the 5th, 1961, with Alan Shepard as the astronaut. It was part of Project Mercury which was an attempt by the USA to bring an astronaut into orbit around the Earth before the Soviet Union during the Cold War. This first manned mission, however, was only a 15-minute suborbital flight, meaning above the limit of space at an altitude of 50 miles and down again. The last part of the mission name came from the Redstone rocket that was used for launching the spacecraft. It was the fourth mission by that rocket in the project, the former being unmanned test flights, one of which carried a chimpanzee. The launch of MR-3 took place at Cape Canaveral, Florida close to the Atlantic Ocean. After the rocket had burned out, the spacecraft with Shepard on board separated from it and continued until it reached an altitude of 116.5 miles (187.5 km) before falling back and landing by parachute on the ocean off of the Bahama Islands. Here it was picked up by helicopter and brought to an aircraft carrier. During the flight, Shepard observed the Earth and tested the reaction control system of the spacecraft together with a pack of small rockets meant for bringing a spacecraft down from orbit on later missions. It was also Shepard who had given the mission its alternative name, "Freedom 7," setting a trend of astronauts naming their spacecraft for the rest of the project.

Friendship 7 … Mercury-Atlas 6 (MA-6) was a human spaceflight mission conducted by NASA. As part of Project Mercury, MA-6 was the successful first attempt by NASA to place an astronaut into orbit. The MA-6 mission was launched February the 20th, 1962. It made three orbits of the Earth, piloted by astronaut John Glenn, who became the first American to orbit the Earth. The event was named an IEEE Milestone in 2011. The Mercury spacecraft, named Friendship 7, was carried into orbit by an Atlas LV-3B launch vehicle lifting off from Launch Complex 14 at Cape Canaveral, Florida. After four hours and 56 minutes in flight the spacecraft re-entered the Earth's atmosphere, splashed down in the Atlantic Ocean and was safely taken aboard the USS Noa. As I sat and wrote about John Glenn and his amazing career, I had tuned into Fox News on Thursday morning, November the the 8th, 2016 and heard the report of his death at 95 years of age, due to complications from cancer. He was, without a doubt an incredible human being. He flew combat missions in World War II and during the Korean War, was the first American to orbit the Earth, was a sitting U.S. Senator from Ohio, flew on the Discovery shuttle as the oldest astronaut on record, and was an inspiration to everyone. He is sorely missed.

Galileo … An orbiting satellite launched from STS-34 to the planet Jupiter; USA.

GAS … STS-57; STS-59; STS-68; USA.

Gemini … Project Gemini was the second human spaceflight program of NASA, the civilian space agency of the United States government. Project Gemini was conducted between projects Mercury and Apollo, with ten manned flights occurring in 1965 and 1966. Its objective was to develop space travel techniques in support of Apollo, which had the goal of landing men on the Moon. Gemini achieved missions long enough for a trip to the Moon and back, perfected extra-vehicular activity (working outside a spacecraft), and orbital maneuvers necessary to achieve rendezvous and docking. All manned Gemini flights were launched from Cape Canaveral, Florida using the Titan II Gemini launch vehicle (GLV).

GRO … Gamma Ray Observatory deployed from STS-37; USA.

HST … Hubble Space Telescope launched from STS-31; USA.

IML-1 … Spacelab International Microgravity Laboratory; STS-42; USA.

IML-2 … Spacelab International Microgravity Laboratory; STS-65; Shuttle Transport System number 65; USA.

Inflatable Antenna Experiment IAE … STS-77; USA.

Intelsat-VI … Recovered from orbit and redeployed from space; STS-49; USA.

ISS … The International Space Station.

JAXA … The Japanese Aerospace Exploration Agency.

Kvant … The Kvant (Russian: Квант-1; English: Quantum-I/1) (37KE) was the first module to be attached in 1987 to the Mir Core Module, which formed the core of the Soviet space station Mir. It remained attached to Mir until the entire space station was deorbited in 2001.

Lacrosse … Imaging radar satellite for all-weather day and night reconnaissance; USA.

LAGEOS 2 … STS-52; USA.

LDEF … Long Duration Exposure Facility satellite, launched from the STS missions; USA.

LEM … Lunar Excursion Module for the Apollo program.

Liberty Bell 7 … Mercury-Redstone 4 was the second United States manned space mission, launched on July the 21st, 1961. The Mercury program suborbital flight used a Redstone rocket. The spacecraft was named Liberty Bell 7 and was piloted by astronaut Virgil I. "Gus"

Grissom. It reached an altitude of more than 118.26 mi (190.32 km) and traveled about 300 mi (480 km). The Redstone was MRLV-8 and the spacecraft was Mercury spacecraft #11, the first with a centerline window instead of two portholes.

LITE … Laser experimentation package; STS-64; USA.

LMS … Life and Microgravity Science Spacelab; STS-78; USA.

Magellan … A satellite launched from STS-30 towards the planet Venus; USA.

Mercury … Project Mercury was the first human spaceflight program of the United States. It ran from 1959 through 1963 with two goals: putting a human in orbit around the Earth, and doing it before the Soviet Union, as part of the early space race. It succeeded in the first but not the second: in the first Mercury mission on May the 5th, 1961, Alan Shepard became the first American in space; John Glenn became the first American to reach orbit on February the 20th, 1962, during the third manned Mercury flight. The program included twenty unmanned launches, followed by two suborbital and four orbital flights with astronaut pilots. Early planning and research were carried out by the National Advisory Committee for Aeronautics (NACA), but the program was officially conducted by its successor organization, NASA. It also absorbed the USAF program Man In Space Soonest which had had the same objectives. Mercury laid the groundwork for Project Gemini and the follow-on Apollo Moon-landing program.

Mir …Mir (Russian: Peace or World) was a space station that operated in low Earth orbit from 1986 to 2001, at first by the Soviet Union and then by Russia. Assembled in orbit from 1986 to 1996, Mir was the first modular space station and had a greater mass than that of any previous spacecraft, holding the record for the largest artificial satellite orbiting the Earth until its deorbit on March the 21st, 2001 (a record now surpassed by the International Space Station). Mir served as a microgravity research laboratory in which crews conducted experiments in plant biology and human biology, physics, astronomy, meteorology and spacecraft systems in order to develop technologies required for the permanent occupation of space.

MMU ... Manned Maneuvering Units (MMU) were used by astronauts conducting EVAs when they "flew" free from the confines of the Shuttle or the ISS.

NASA ... The National Aeronautical and Space Agency; USA.

NIH-Experiments ... STS-59; USA.

OARE ... STS-75; USA.

OAST ... The Office of Aeronautics and Space Technology (OAST) platforms carried aloft by the Shuttle fleet for experiments on board the ISS.

OAST-2 ... STS-62; USA.

ORFEUS-SPAS ... STS-51; USA.

PAMS ... STS-77; USA.

Priroda ... A Russian segment for the ISS, used for experiments in weightlessness.

Progress ... A Russian unmanned resupply vessel; USSR.

R-5A ... An intermediate range Soviet missile converted to put into space on sub-orbital flights a human-capable capsule. All the R-5A flights failed with the death of several cosmonauts.

RMS ... Canadian designed robotic arm system used first on board the U.S. space shuttles and then on the International Space Station (ISS); USA.

ROMPS ... STS-64; USA.

Rossiya ... The name given to Vladimir Ilyushin's Vostok space capsule. Ilyushin was the first person to survive space flight. His flight date was April the 7th, 1961.

RSA ... The Russian Federal Space Agency.

SAFER (EVA) ... STS-64; USA.

Salyut ... The Salyut program (Russian: Salute or Fireworks) was the first space station program undertaken by the Soviet Union, which consisted of a series of four crewed scientific research space stations and two crewed military reconnaissance space stations over a period of 15 years from 1971 to 1986. It was, on the one hand, designed to carry out long-term research into the problems of living in space and a variety of astronomical, biological and Earth-resources experiments, and on the other hand this civilian program was used as a cover for the highly secretive military Almaz stations, which flew as well under the Salyut designation.

SAREX-II … STS-78; USA.

SFU Space Flyer Unit … STS-72; USA.

Shuttle … The Space Shuttle was a crewed, reusable Earth orbital spacecraft operated by NASA. Its official program name was Space Transportation System, taken from a 1969 plan for a system of reusable spacecraft of which it was the only item to be funded for development. The first of four orbital test flights occurred in 1981, leading to operational flights beginning in 1982. It was used on a total of 135 missions from 1981 to 2011, all launched from Cape Canaveral in Florida. Major missions included launching numerous satellites, interplanetary probes, the Hubble Space Telescope (HST), conducting space science experiments, and constructing and servicing the International Space Station. Major components included the orbiters, recoverable boosters, the external tanks, payloads, and supporting infrastructure. Five space-worthy orbiters were built; two were lost in mission accidents.

The Space Shuttle at launch consisted of the Orbiter Vehicle (OV), one external tank (ET), and two Solid Rocket Boosters (SRBs). It was launched vertically like a conventional rocket with thrust from the two SRBs and three main engines. During launch, the external tank provided fuel for the orbiter's main engines. The SRBs and ET were jettisoned before the orbiter reached orbit. At the conclusion of the orbiter's space mission, it fired its thrusters to drop out of orbit and re-enter the lower atmosphere. The orbiter decelerated in the atmosphere before flying like a glider but with reaction control system thrusters before landing on a long runway. Columbia, Challenger, Discovery,

Atlantis, and Endeavour were the space-capable orbiters that were built.

Sigma 7 ... Mercury-Atlas 8 (MA-8) was the fifth United States manned space mission, part of NASA's Mercury program. Astronaut Walter M. Schirra, Jr., orbited the Earth six times in the Sigma 7 spacecraft on October the 3rd, 1962, in a nine-hour flight focused mainly on technical evaluation rather than on scientific experimentation.

Skylab ... Skylab was a space station launched and operated by NASA and was the U.S.'s first space station. Skylab orbited the Earth from 1973 to 1979, and included a workshop, a solar observatory, and other systems. It was launched unmanned by a modified Saturn V rocket, with a mass of 169,950 pounds (77 tons). There were three manned missions to the station, conducted between 1973 and 1974 using the Apollo Command/Service Module (CSM) atop the smaller Saturn 1B; each delivered a three-astronaut crew. On the last two manned missions, an additional Apollo / Saturn IB stood by ready to rescue the crew in orbit if it was needed.

SL-J ... Japanese Spacelab; STS-47: Japan and the USA.

SLS-1 ... Spacelab for Life Science; STS-40; USA.

SLS-2 ... Spacelab for Life Science; STS-58; USA.

SLR ... Space Radar laboratory for the International Space Station, carried aloft by the American Shuttle fleet.

SLR-2 ... Space Radar laboratory; STS-68; USA.

SMM ... The Solar Maximum Mission satellite (or SolarMax) was designed to investigate Solar phenomena, particularly solar flares. It was launched on February the 14th, 1980. Although not unique in this endeavor, the SMM was notable in that its useful life, when compared with similar spacecraft, was significantly increased by the direct intervention of a manned space mission. During STS-41-C in 1984, the Space Shuttle Challenger intercepted the SMM, maneuvering it into the shuttle's payload bay for maintenance and repairs. SMM had been fitted with a shuttle grapple fixture so that the shuttle's robot arm could grab it for repair.

SSBUV/A … STS-62; USA.

Soyuz … Soyuz (Russian: Union) is a series of spacecraft initially designed for the Soviet space program by the Korolyov Design Bureau in the 1960s, which is still in service today. The Soyuz succeeded the Voskhod spacecraft and was originally built as part of the Soviet manned Lunar program.

Spacehab-3 … STS-63; USA.

Spacehab-4 … STS-77; USA.

Space Radar Lab … Mapped the Earth's surface in three dimensions; STS-59; USA.

SPAS-1 … American military strategic defense satellite, with six launches, June the 18th, 1983 (SPAS-01) to July the 8th, 1997 (CRISTA). The SPAS (Shuttle Pallet Satellite) satellite was a reusable free-flying vehicle built by Messerschmitt-Bolkow-Blohm, which could be deployed and then retrieved by the U.S. Space Shuttle's Remote Manipulator System arm. The original SPAS, with materials processing and SDI-related sensor payloads, was used on several missions (STS-7, STS-11, STS-39). An experiment-carrying truss (USS) based on the original SPAS structure (but without the avionics and attitude control) was flown on the Spacelab D-1 and D-2 missions.

SPARTAN … STS-72; STS-77; USA.

SPARTAN-2 … STS-56; USA.

SPARTAN-4 … STS-63; USA.

SPARTAN/OAST Flyer … STS-72; USA.

Spektr … The Spektr project is funded by the Astro Space Center of Russia, and was launched into Earth orbit on July the 18th, 2011, with a perigee of 10,000 kilometers (6,200 mi) and an apogee of 390,000 kilometers (240,000 mi), about 700 times the orbital height of the Hubble Space Telescope. The main scientific goal of the mission is the study of astronomical objects with an angular resolution up to a few millionths of an arcsecond. This is accomplished by using the satellite in conjunction with ground-based observatories and interferometry techniques.

Stamps … STS-68; USA.

Syncom satellite … Syncom (synchronous communication satellite) started as a 1961 NASA program for active geosynchronous communication satellites, all of which were developed and manufactured by Hughes Space and Communications. Syncom 2, launched in 1963, was the world's first geosynchronous communications satellite. Syncom 3, launched in 1964, was the world's first geostationary satellite. In the 1980s, the series was continued as Syncom IV with some much larger satellites, also manufactured by Hughes. They were leased to the United States military under the Leasat program.

Taikonaut … A Chinese astronaut.

TDRS … Tracking and Data Recovery Satellite; USA.

TDRS-E … Tracking and Data Relay Satellite; deployed from STS-43; USA.

TDRS-F … Tracking and Data Relay Satellite; deployed from STS-54; USA.

TDRS-G … Tracking and Data Relay Satellite; deployed from STS-70; USA.

TEAMS … STS-77; USA.

TSS-1 … Tether experiment package; STS-46; USA.

TSS-1R … Tether Satellite: STS-75; USA.

UARS … Upper Atmosphere Research Satellite; deployed from STS-48; USA.

Ulysses … The Ulysses Solar Probe launched from STS-41; USA.

USML-1 … U.S. Microgravity Laboratory; STS-50; USA.

USML-2 … U.S. Microgravity Laboratory; STS-73; USA.

USMP … U.S. Microgravity Payload; STS-52; USA.

USMP-2 … U.S. Microgravity Payload; STS-62; USA.

USMP-3 … U.S. Microgravity Payload; STS-75; USA.

UV Astronomy Spacelab … STS-67; USA.

VKA Myasishchev … An early Soviet-designed space plane that was used once in a failed attempt to orbit a female cosmonaut. She died during the attempt.

Vostok … The Vostok (Russian: Восток, translated as East) was a type of spacecraft built by the Soviet Union.

Voskhod … The Voskhod (Russian: Sunrise) was a manned Soviet spacecraft. It was the first spacecraft to carry more than one crewman into orbit, the first flight without the use of spacesuits, and the first to carry either an engineer or a physician into outer space.

Wake Shield facility … U.S. experiment for creating artificial vacuums in space; STS-60; STS-69; USA.

X-15 … The North American X-15 was a rocket-powered aircraft operated by the United States Air Force and NASA as part of the X-plane series of experimental aircraft. The X-15 set speed and altitude records in the early 1960s, reaching the edge of outer space and returning with valuable data used in aircraft and spacecraft design. As of this date, notwithstanding potential secret Air Force or CIA-related projects that I have no knowledge of, the X-15 holds the official world record for the fastest speed ever reached by a manned aircraft. Its maximum speed was 4,520 miles per hour (7,274 km/h).

Acknowledgements

Sergei Khrushchev … the interview took place at Brown University in Providence, Rhode Island on Tuesday, May the 4th, 1993, at 10:30a.m..

Jaden Yaboz … in Haifa, Israel … several in-person interviews and phone conversations … August of 2016; December of 2016; April of 2017; December of 2017; June of 2018; January of 2019; October the 14th of 2020; March of 2021; October the 14th of 2021; February the 4th of 2022.

Giovanni Battista … of the Italian police … the interview took place in Florence, Italy in May of 1997.

Dr. James Lawrence … NASA scientist … the interview took place in Langley, Virginia on October the 19th, 1992 at 3:00p.m. … he passed away on April the 30th, 2002.

Malikova Tatiana Yurivna … the telephone conversations/interviews from Moscow took place in June of 2018, and in February of 2019.

James Oberg … information from his book was used … "Uncovering Soviet Disasters – Exploring the Limits of Glasnost" (Random House, New York; 1988; PN5277. D58024; pages 162 to 163).

Charlie Adams … of NASA … radio host on-air statements from 1997.

Pravda news release concerning Yuri Gagarin on April the 12th, 2001.

Television documentary entitled "Cosmonaut Cover-Up" appeared on the UK's Horizon Channel in 2001.

Dennis Ogden of the UK's "The Daily Worker" article in 2001 … and "Yuri Gagarin Conspiracy: Fallen Idol" 2009.

Lord Bruce-Gardyne of the UK's "British Daily Telegraph" article in 2001.

Mary Bennett's and David Percy's book, "Dark Moon – And The Whistle Blowers" writings … April the 2nd, 2001.

The Russian newspaper, "Krasnaya Zvezda" article … January the 29th, 1983.

General Nikolai Kamanin … former second in command of the cosmonauts … notes that were confided in his diary of April, 1961.

Aleksander Zheleznyakov … taken from articles for the "Orbit And Spaceflight" journals (Orbit, Journal of the Astro Space Stamp Society; issue number 54, June of 2002; Spaceflight, Volume 44, Number 11, November of 2002, pages 471 to 475).

Dr, Chris Riley … in the UK … (2007 "In The Shadow of the Moon"; 2008 "When We Left Earth" - The NASA Missions from the Discovery Channel; 2009 - "One Small Step" from the BBC Worldwide).

Notes from the 2007 documentary in Italy, entitled, "I Pirati dello Spazio (Pirates of Space).

Notes from the science magazine, "Fortean Times" … March of 2008.

The BBC's radio broadcast, "Listen Up" by Glen Neath on Radio 4 in England, May of 2009.

The television science channels show, "Dark Matters: Twisted But True" in 2011.

Notes from Julius Epstein's in-depth article, for the Houston Chronicle" published on October the 5th, 1967.

Notes from Frank J. Gaffney's article in the March/April 2022 edition of of the Association of Mature American Citizens (AMAC) magazine.

Notes from Dr. Laura A. Whitlock of NASA's Star Child project, "All-Sky Monitor Database Update" … (Legacy, Number 4, 1994).

Index

1997. [TV programme] WKVI K99.3 FM: Charlie Adams.

Battista, G., 1997. .

Bennett, M. and Percy, D., 2001. *Dark moon*. Kempton, Ill.: Adventures Unlimited Press.

Bruce-Gardyne, L., 2001. British Daily Telegraph.

Cordiglia, A., n.d. .

Cosmonaut Cover-Up. 2001. [video] UK's Horizon Channel.

Dark Matters: Twisted But True. 2011. [video] Directed by D. Gold and J. James. Wide-Eyed Entertainment.

Elliott, H., n.d. .

Epstein, J., 2022. *The Houston Chronicle,*.

Epstein, J., n.d. .

Fortean Times, 2008. .

Gaffney, F., 2022. *Association of Mature American Citizens (AMAC),*.

I Pirati dello Spazio. 2007. [film] Italy.

In The Shadow of the Moon. 2007. [film] Directed by D. Sington and D. Riley. Film4 Passion Pictures Discovery Films.

Khrushchev, S., 1993. .

Krasnaya Zvezda, 1983. .

Lawrence, D., 1992. .

Listen Up, 2022. [TV programme] 4: Glen Neath.

Oberg, J., 1988. *Uncovering Soviet Disasters; Exploring the Limits of Glasnost*. New York, New York, Random House: Random House, New York, pp.162 - 163.

Ogden, D., 2001. The Daily Worker.

Ogden, D., 2009. Yuri Gagarin Conspiracy: Fallen Idol.

One Small Step. 2009. [film] Directed by D. Cassidy and K. Davy. Taiko Studios.

When We Left Earth: The NASA Missions. 2008. [film] Directed by D. Riley. Discovery.

Whitlock, D., 2022. *All-Sky Monitor Database Update*. [online] Heasarc.gsfc.nasa.gov. Available at: <https://heasarc.gsfc.nasa.gov/docs/journal/asmdb4.html> [Accessed 20 April 2022].

Yaboz, J., 2022. .

Yurivna, M., 2019. .

Zheleznyakov, A., 2002. Orbit And Spaceflight. *Journal of the Astro Space Stamp Society*, 44(54), pp.471- 475.

Author's Notes

A personal note of appreciation goes out to a number of people who helped with this book in one fashion or another. Sometimes it was a word of encouragement or some timely advice, but, whatever it was, the help they offered was so very deeply appreciated. Their names are as follows:

First of all, I must thank my two amazing and wonderful daughters, Mia and Illianna (Illy), for without them in my life, the future would have no meaning. My irreplaceable father, Henry H. Elliott, Jr., who convinced me long ago to push forward with this project. He is missed so very much.

My tireless editor, Jaden Yaboz, still flying by the seat of his pants out of Haifa, Israel. And, of course, my wife, Deb A. Watts Elliott, who is without a doubt one of the best proofreaders and editors in the world!

My agent, social media expert, publicist extraordinaire, wonderful friend, and advocate for all of my projects, Lauri Stevens.

I also must thank NASA's Charlie Adams, the untimely late Professor Sergei Khrushchev, Dr. James Lawrence, Catherine Martin, Mikhail Rudenko, Laura Whitlock of NASA's Star Child Project, and Malikova Tatiana Yurivna, the Head of the Department of Complex Simulators for the Russian segment of the ISS (RSS-ISS), who helped with some of the information and many of the translations.

About the Author

John Elliott is the author of twenty-seven published works, and a writer for the *Chicken Soup for the Soul* series of best-selling books. He also writes articles for a number of on-line concerns, and previously wrote book reviews and conducted editorial work for the National Audubon Society's and the Smithsonian Institute's *Northeastern Naturalist* and *Southeastern Naturalist* through the Humboldt Institute of Maine. He is also a member of the Society of Children's Book Writers and Illustrators. In addition to that, he has written numerous articles for the United States Department of Justice quarterly review, *The Thin Blue Line*. He is the owner of and editor for *Chappy's Gun World Writers*, and a columnist for the *Australian Police Journal*. Hollywood is also in the process of creating a full length motion picture based on his prior best-selling autobiography, *The August Assassin,* and the British Broadcasting Corporation in the United Kingdom is creating a television situation comedy based on the book about the his father, entitled *Life According to Dad*. He was an on-air contributor for the BBC's *Sunday Morning Live* shows in London, England, and in Belfast, Northern Ireland, and a commentator for Radio Zabok in Croatia.

He is a retired forty-four year law enforcement veteran, having worked for various police agencies in the United States, as well as for Interpol in Lyon, France, and conducted joint investigations with Europol. For many of those years he worked concurrently on a contractual basis for the Special Operations Group of the Central Intelligence Agency, as well as for the Israeli spy agency Mossad. Fluent in English, Gaelic, Hebrew and Hungarian, he can also converse in both French and Italian. He holds a Bachelor of Science degree in business, an MBA, and a Juris Doctorate law degree. As a public and motivational speaker, he conducts safety seminars nationwide. He is also an outspoken advocate for America's law enforcement officers and military personnel, and has trained countless police recruits in the United States and abroad. He can be directly contacted through his website www.JohnElliottBooks.net.

Also Available by John Elliott

The August Assassin

Zephyr

Life According to Dad

Living With Henry

Don't Be a Victim! An Officer's Guide on Preventing Crime

Surviving Terrorism

Angel Interactions and the Words of God's Servants

The Other Side of Suicide

Savages – The Newest Old Threat to World Peace

Rescuing Rascal

Ri Ra – An Adventure Begins

After The Eleventh Hour

God for Men

David's Stone – How to Live Without Fear

The Vatican Betrayals

Sara and Harris

The Corgi and The Cairn Dragon Feast

The Ramblings of a Law-Abiding Gun Nut

The Bedtime Storyteller

Something Strange Happened on the Way to the Police Station

The Ellsworth Express

Dad's Daughters – You Have to Get it Right The First Time

Warpaint

Chicken Soup for the Soul - several *New York Times* Best Selling
series of releases

Australian Police Journal - on-going monthly releases

Chappy's Gun World Writers - on-going weekly releases

The Thin Blue Line

Northeastern Naturalist

Southeastern Naturalist

Soviet Space Disasters – A Nation's Hidden Disgrace Revealed